THE ROLE OF

MATHEMATICS

IN THE RISE

OF SCIENCE

THE ROLE OF

MATHEMATICS

IN THE RISE OF

SCIENCE

BY SALOMON BOCHNER

PRINCETON, NEW JERSEY

PRINCETON UNIVERSITY PRESS

Published by Princeton University Press,
Princeton, New Jersey
In the United Kingdom: Princeton University Press,
Guildford, Surrey
Copyright © 1966 by Princeton University Press
ALL RIGHTS RESERVED
L.C. Card: 66-10550
ISBN: 0-691-08028-3
ISBN: 0-691-02371-9 pbk.
Printed in the United States of America by
Princeton University Press at Princeton, New Jersey
Fourth Printing, 1981

First Princeton Paperback printing, 1981

PREFACE

THE CHAPTERS of this book were originally separate essays for several purposes, and all but one have been published before, beginning in 1961. The essays have now been revised, and linked up; and an Introduction, and a collection of Biographical Sketches in lieu of an Epilogue, have been added.

This is, throughout, a book about mathematics, even if in various contexts it is seemingly concerned with other matters. It is a book about the uniqueness of mathematics as a force of our intellectuality, and about the mystique of its creativity; about the growing efficacy of mathematics, its widening importance, and its continuing spread. It is intended to be a book not *in* mathematics but *about* mathematics, even if some parts of it, which do not affect the rest of the book, are involved with outright mathematical technicalities; but no attempt is made to forcibly translate snippets or samples of technical mathematics into non-technical vernacular.

Ours is an age in which scientists are Wise Men, and the root of this Wisdom is in Mathematics. But mathematics, if taken by itself, is almost a pastime only, albeit an esoteric one; or so it seems. What makes mathematics so effective when it enters science is a mystery of mysteries, and the present book wants to achieve no more than to explicate how deep this mystery is.

Mathematics is both young and old; and in order to comprehend its role in our Knowledge of today, it is fruitful, and even imperative, to understand the portends of the vicissitudes of the mathematics of yesterday. Thus, we will frequently turn, for comparison and contrast, to Greek mathematics, and also Greek physics, and even to the enveloping Greek Rationality of which they are token and texture.

[v]

It gives me pleasure to record that after the first essays of this collection began to appear in periodicals, Mr. Charles Scribner, Jr., president of Princeton University Press, noticing them, soon suggested that they be made into a book.

August, 1965   S. B.

CONTENTS

[vii]

THE ROLE OF
MATHEMATICS
IN THE RISE
OF SCIENCE

INTRODUCTION

THE ESSAYS of this collection are all concerned with the role of mathematics in the rise and unfolding of Western intellectuality, with the sources and manifestations of the clarity and the mystique of mathematics, and with its ubiquity, universality, and indispensability.

We will frequently confront the mathematics of today with the mathematics of the Greeks; and in such a confrontation it is pertinent to take the entire mathematical development since A.D. 1600 as one unit. Therefore, "modern mathematics" will mean for us, invariably, mathematics since 1600, and not since some later date, even though, for good reasons, the mathematics of most of the 20th century is clamoring for an identity and definition of its own.

For our retrospections, the sobering fact that Greek mathematics, the mighty one, eventually died out in its own phase is of much greater import than the glamorous fact that Greek mathematics had come into being at all, although historians rather exult in its birth than grieve at its death. T. L. Heath in his standard history of Greek mathematics [1] finds it awe-inspiring to contemplate how much Greek mathematics achieved "in an almost incredibly short time"; and the context makes it clear that Heath means the time from 600 B.C., when the first Thales-like geometry began to stir, to 200 B.C., when Apollonius wrote his treatise on *Conics*. This makes 400 long years, and pronouncements like that of Heath are insultingly condescending to the Greeks, especially painfully so since immediately after the *Conics,* and thus 50 years after the *floruit* of Archimedes, Greek mathematics was at a loss where to turn next. In irreconcilable contrast to this, in the 400 years since A.D. 1565 mathematical intellectuality has turned the world topsy-turvy many times over, and shows no sign of abating.

[1] T. L. Heath, *A History of Greek Mathematics* (1921), I, 1.

Any major intellectual growth has to have, from time to time, bursts of unconscionably fast developments and stages of dizzying precipitousness, unless the developments are in anthropologically early phases of pre-intellectuality only. Much more "incredible" than the development of Greek mathematics in the course of 400 years was the reorientation of 20th-century physics, by a handful of youngsters, in the course of the four years from 1925 to 1928. Or, if to stay within Greek antiquity, nothing in the development of Greek mathematics can match, in sheer speed, Aristotle's amassment and *articulation* of logical, metaphysical, physical, cosmological, biological, and sociological knowledge within at most 40 years of his life (he died at the age of 62). As regards the growth of intellectuality, the plea of G. B. Shaw that man's life span ought to be at least 300 years [2] is of dubious merit.

Greek mathematics, whatever its inspiredness and universality, was slow, awkward, clumsy, bungling, and somehow sterile; and the limitations of Greek mathematics have wide implications. By its nature and circumstances Greek mathematics was part of Greek philosophy, that is, of the philosophies of Parmenides, Plato, and Aristotle, and it thus was a faithful image of a large segment of Greek intellectuality as a whole. Therefore, weaknesses of the mathematics of the Greeks were weaknesses of their intellectuality as such; and just as the death of the Great Pan signified the end of Greece's chthonic vitality,[3] so also the death of the mathematics of Archimedes signified, perforce, the end of Olympian [4] intellectuality.

[2] In the Preface to *Back to Methuselah*.

[3] Jane Harrison, *Prolegomena to the Study of Greek Religion*, 3rd ed. (1922), also reprinted as Meridian Book (1957), p. 651; Archer Taylor, "Northern Parallels to the Death of Pan," *Washington University (St. Louis) Studies*, Vol. 10, Part 2, No. 1 (October 1922), p. 3.

[4] Our distinction between "chthoric" and "Olympian" is that of the work of Jane Harrison, n. 3.

Yet the Greeks had extraordinary anticipations of the role of mathematics. It is an attribute of our times that mathematics is growing differently from other disciplines. All areas of knowledge are growing by expansion from within and accretion from without, by internal subdivision and external overlappings with neighboring areas. Mathematics, in addition to that, is also penetrating into other areas of knowledge one-sidedly, for their benefit, and more by invitations and urgings of the recipient fields than from an expansionist drive. Now, a glimmer of this peculiar and unique universality of mathematics was noticed, however dimly and inarticulately, by the "typically" Greek mathematics, even when still in the Pythagorean swaddling clothes of its infancy.[5] Aristotle repeatedly states that Pythagoreans were saying that all things are numbers, that substance (that is, matter in our sense) is number, etc.; now, this has such a modern ring that it could be lifted out of a book on elementary particles. Aristotle also reports that the Pythagoreans found that acoustical "harmonies" only depend on (commensurate) ratios of lengths and on nothing else, and that they generalized from this experience in acoustics to physics and knowledge in general. This was an uncanny anticipation; the "linear oscillator" is a simple device of acoustical provenance, but it also is an elemental skeletal part of theoretical physics of today. It is true that, against their own background, the Pythagoreans were recklessly imprudent when concluding so much from so little. But Isaac Newton was also "reckless" when proposing that there is a universal gravitational law which subsumes both terrestial falls of bodies and celestial orbitings of planets. In fact, a first confirmation of Newton's law came only in the 20th century when man-made satellites did indeed move according to the law of Newton; until then the law was a "feigned" hypothesis, all Newton's protestations notwith-

[5] A satisfactory recent report on Pythagoreans is in W. K. C. Guthrie, *A History of Greek Philosophy* (1962).

standing. Finally, it is undoubtedly true that the Pythagoreans drew their conclusions about the role of certain commensurate ratios in physics when they were still under the ingenuous misconception that all ratios are automatically commensurate, that is, before they discovered the existence of noncommensurate ratios of ordinary lengths. But this is of no consequence; Planck, Einstein, Niels Bohr, etc., knew that there are irrational numbers, but they laid the foundation of quantum theory on a groundwork of integer numbers and commensurate ratios nonetheless.

There is a common thesis that due to political, technological, sociological, and similar inadequacies the entire fabric of Hellenistic civilization gradually disintegrated and that the decline of mathematics was a part of this decline in general. If one accepts this thesis then one can even argue that, due to its greater strength, mathematics withstood the onslaught of the decline longer than other areas of intellectuality. In fact, Greek tragedy did not maintain itself on the level of Aeschylus, Sophocles, and Euripides beyond 400 B.C., nor did Greek history maintain itself on the level of Herodotus and Thucydides beyond that date. Greek philosophy did maintain itself on the level of Parmenides and Socrates beyond 400 B.C., through Plato and Aristotle, but only till 322 B.C. when Aristotle died. However, just then, Greek mathematics, which had begun in the 6th century B.C., made an extraordinary exertion and maintained itself on a high level for another 150 years or so. Indeed, one might even say that this post-Aristotelian phase was the true Golden Age of Greek mathematics, if a golden age there must be. In fact, sometime between 322 and 300 B.C. Euclid wrote his *Elements,* which was the greatest "primer" in anything, ever; the middle of the 3rd century B.C. was the age of Archimedes, in whose style Newton still composed his *Principia*; and around 200 B.C.

Apollonius wrote his *Conics*, with the aid of which, 1,800 years later, Kepler finally smashed the confining planetary circle which Plato had proclaimed to be divinely beautiful and eternal.

But if, granting all this, one nevertheless persists and asks the very pointed question how it came about that Greek mathematics immediately after the *Conics*, when still at its height, and seemingly at its healthiest, suddenly began to falter, then no explanation from general causes is available.

Explanations from sociological causations are also unavailable for problems of modern mathematics. For instance, it seems impossible to explain from sociology why in the 18th century and only then, when there was little industrial inducement and very little attention to experimental verification, there was a near-perfect, and richly yielding, fusion between mathematics and mechanics, to the immeasurable advantage of both, and of all physics and technology to come; whereas in the 19th century, under the Argus-like stare of advantage-seeking industrial capitalism, a so-called Applied Mathematics was beginning to splinter off, about which it cannot be said even today whether it was necessary or fortuitous, a good thing or a bad thing. And it is not easy to gauge what the eventual impact of the presently proliferating mathematics-to-various-purposes will be. The proliferation might even recede; there is a precedent for this. At the turn of the century there was a widespread movement for the propagation of Hamiltonian quaternions, on the grounds, apparently, that it is a most applicable technique. The movement has long since been extinct.

Most of the growing mathematics-to-a-purpose will probably endure. But we wish to say that the "purer" mathematics is, the more it embodies the significant designs of the texture of knowledge; for this reason, in the past, more significant applications to basic science came from mathematics that had been pursued for its own sake than from mathematics that had been pursued to a purpose.

A heavy sociological accent rests on the contemporary movement, especially in the United States, for reforming the mathematical curriculum in pre-college schools. The movement will undoubtedly prevail; but there may be an "disadvantaged generation" of youngsters who will have to pay a transition price, before a certain retardation in the learning of basic "operational" skills, which seems to be incident to the reform, is minimized. The last such reform was, at the turn of the century, Felix Klein's introduction of the infinitesimal calculus into the final stages of school training, and this reform began to erase the distinction between lower and higher mathematics which had been institutionalized in the course of the 19th century. The present-day movement aims, in substance, at introducing into school training Georg Cantor's set theory, and at an early stage, too. The tempo of developments quickens; the calculus entered schools after 200 years, set theory after 75 years. Cantor's set theory is a fountainhead of modernism-by-universality, or of universality-by-modernism, in 20th-century mathematics; the theory sets in at precisely the point at which mathematics proper sets off from rationality in general. Cantor well knew what his mission was; his memoirs have, in their footnotes, pertinent references to Plato, Aristotle, and other ancients, and to earlier figures in modern mathematics.[6] He asserts that he is succeeding where others have failed; but there is an undertone of awareness that he is succeeding because there were others who tried before him, even if they failed. And Cantor's attention to "ancient history," when himself on the height of achievements for the future, might be viewed as a lesson too.

[6] Georg Cantor, *Gesammelte Abhandlungen* [Collected Works], edited by E. Zermelo (1932).

A further problem with a sociological accent, more a problem of the future than of the past, and perhaps more interesting for mathematics than for other science, is the question of both metaphysical and "practical" import: whether originality of creativity resides, as is commonly taken for granted, decidedly in the individual, without being able to become corporate and pluralistic or whether originality of creativity might not be able to manifest itself in the achievements of a group of individuals jointly, without meaningful distribution over the component individuals severally.

Since the 1940's there have been, all over the world, ever more research efforts through "programs," "projects," and other concerted groupings, and it is difficult to foretell the meaningfulness of all this. It would cetainly be rash to assert that all such groupings are only outward settings without inward consequences, and that it will always be only the individual in his separateness who will count. All of science, and in fact all of organized knowledge is much too "recent" an adventure of *Homo sapiens* for suchlike assertions to be made. It may turn out that amidst the "lonely crowd" of tomorrow research activity will become pluralistic, even if in the "gregarious crowd" of yesterday it was all individualistic; that the originality of the brief yesterday will become the routine of the long tomorrow; and that all the philosophemes about all our knowledge thus far will have been no more than the uneasy dream between dawn and awakening.

PART I. ESSAYS

CHAPTER 1

FROM MYTH TO MATHEMATICS
TO KNOWLEDGE *

1. WHAT IS MATHEMATICS?

WHAT INDEED is mathematics? This question, if asked in earnest, has no answer, not a satisfactory one; only part answers and observations can be attempted.

A neat little answer, a citizen's description of mathematics in capsule form, is preserved in the writings of a Church Father of the 3rd century A.D.; Anatolius of Alexandria, bishop of Laodicea, reports that a certain (unnamed) "jokester," using words of Homer which had been intended for something entirely different, put it thus: [1]

Small at her birth, but rising every hour,
While scarce the skies her horrid [mighty] head can bound,
She stalks on earth and shakes the world around.
(*Iliad,* IV, 442–443, Pope's translation.)

For, explains Anatolius, mathematics begins with a point and a line, and forthwith it takes in heaven itself and all things within its compass. If the bishop were among us today, he might have worded the same explanation thus:

* Originally in *Journal of the History of Ideas,* Vol. 26 (1965), pp. 3–24, under the title "Why Mathematics Grows." Enlarged.

[1] *The Ante-Nicene Fathers,* edited by Alexander Roberts and James Donaldson, VI, 152.

The leading edition of the Greek text of Anatolius is due to F. Hultsch. This text, which is only a surviving fragment, is included in the collection *Anonymi variae collectiones* which Hultsch appended to his edition of the mathematical works which go under the name of Heron (Berlin, 1864). For bibliography about Anatolius see George Sarton, *Introduction to the History of Science,* I (1927), 337.

For mathematics, as a means of articulation and theoretization of physics, spans the universe, all the way from the smallest elementary particle to the largest galaxy at the rim of the cosmos.

The gist of this and similar explanations is a declaratory statement or assertion, with some illustrations, that mathematics is distinctive and effective, and thus important. But how have these traits of distinctiveness and effectiveness come about, and how do they sustain themselves so powerfully? And whence comes the urge, the intellectual one, to spread mathematics into ever more areas of knowledge, old and new, especially new ones? Perhaps such questions cannot be answered at all. But we will try, and our approach will be analytical and evolutional, both.

2. MATHEMATICS AND MYTHS

We will proceed from a quotation:

> Mathematics is a form of poetry which transcends poetry in that it proclaims a truth; a form of reasoning which transcends reasoning in that it wants to bring about the truth it proclaims; a form of action, of ritual behavior, which does not find fulfillment in the act but must proclaim and elaborate a poetic form of truth.

This appealing sentence is not my own; I wish it were. It is taken from a book on the awakening of intellectuality in Egypt and Mesopotamia, the two near-Mediterranean areas in which, by Chance or Providence, inexplicably but unmistakably, our present-day "Western" civilization germinated first.[2] However, in the book from which it is

---

[2] See the book of Henri Frankfort and others, *The Intellectual Adventures of Ancient Man* (Chicago, 1946), p. 8; also in a Penguin edition, under the title, *Before Philosophy*, p. 16.

Another statement from this book which is meaningful for mathematics instead of myth is the following one from page 12 (or page 21 in the Penguin edition): ". . . there is coalescence of the symbol

taken, the sentence is a pronouncement not on mathematics but on myth. That is, in the original the sentence runs as follows:

> Myth is a form of poetry which transcends poetry in that it proclaims a truth; a form of reasoning which transcends reasoning. . . .

Also, the book from which we quote does not concern itself with mathematics at all; but it might have done so, because in Egypt and Mesopotamia, severally, mathematics started very early, surprisingly so. For instance, it started there about 2,000 years before the famed "classical" mathematics of the Greeks, which was a summit achievement of their natural philosophy, began to shape itself, in its separateness, in the 6th and 5th centuries B.C. It is true that in aspects of intellectuality Greek mathematics was greatly superior to whatever earlier mathematics it built on. But it is also a fact that, by evolutionary precedence, the two mathematics of Egypt and Mesopotamia were firsts, unqualifiedly so, not as anthropological phenomena of a prehistoric man, but as organized systematic pursuits

---

and of what it signifies, as there is coalescence of two objects compared so that one may stand for the other."

Mathematics is also tending, *intentionally,* to coalesce the symbol of something with the something symbolized, and to coalesce two objects of separate provenance if they can operationally replace one another.

Another book of H. Frankfort, *The Birth of Civilization in the Near East* (London, 1951) deals with the vexing problem of prehistoric direct contacts between Mesopotamia and Egypt, the two original foci of Western Civilization. About this see also, for example, Arnold J. Toynbee, *A Study in History,* Vol. XII, *Reconsiderations* (1961), pp. 345–348, and W. H. McNeill, *The Rise of the West* (Chicago, 1962). The latter work is an attempt, of which there have been many beginning with the Old Testament, of presenting a coherent concatenation of circumstances, phenomena, and developments which eventuated in the inception and formation of Western Civilization.

of the *Homo sapiens* who initiated our present-day civilization in its elaborateness.

It is a significant characteristic of Western civilization as a whole that mathematics, however esoteric it may appear to be, was very nearly the first organized Rational Knowledge at all. Early achievements in mathematics are at least as "surprising" as are early achievements in other areas, such as in architecture, visual or musical arts, commerce and voyaging, poetry, and religion and morality.

In Egypt and Mesopotamia, mathematics, although quite old, was not as old as myth-making, and it could not have been so, because mathematics, the "organized" one, presupposes some kind of "organized" writing, which active myth-making probably does not presuppose. But, in Egypt and Mesopotamia, "indigenous" mathematics began not very long after "indigenous" writing had begun; whereas in Greece, the distinctively Greek mathematics began, apparently, well after Greek (poetic) "literacy" had been flowering. And in the land of the Bible, whose literacy has given us most of our moral imperatives, a mathematics of its own never came into being at all.

Plato (427–348 B.C.) was attracted to mathematics, and, seemingly independently from this, he was a master of myth-making. Yet, in the *Timaeus,* which is Plato's "monograph" on cosmogony and cosmology, there is a meaningful mixture of mathematics and myth-making, and Aristotle (384–322 B.C.), when studying the *Timaeus,* took it for granted that even outright mythological features, in addition to being literary "embellishments," ought to have a "scientific" interpretation too. Some commentators, since antiquity, have criticized Aristotle for seeking, unimaginatively, such an interpretation in every detail of Plato's flight of mythological inspiration; but I think that, in retrospect, Aristotle can be justified to an extent. In our contemporary works on cosmology the setting is not mythological, but invariably mathematical. And scientists also demand—

and this has become a part of the celebrated "scientific prediction"—that if in the setting a purely mathematical conclusion arises which had not been expected, then a corresponding occurrence "in nature" which had not been known before shall be detectable. I further think that in Plato's intellectual perception, and then also in Aristotle's, there was, indistinctly, a certain similarity between mathematics and myth-making by which the parallelism of their occurrence becomes justified.

The similarity between mathematics and myths is grounded in a certain similarity of articulation; mathematics and myths both speak in "symbols," recognizably so, and the fact that it is nearly impossible to say satisfactorily, in or out of mathematics, what a symbol is [3] does not destroy the similarity itself. Symbols in myths are very different from symbols in mathematics, but they are symbols all the same.

Symbols in myths, as in poetry, may be charged, intentionally or half-intentionally, with ambivalences, whereas in mathematics they must not so be; nevertheless, even in myths symbols contribute a clarity and incisiveness which is peculiar to them, wherever they occur. Above all, even in myths symbolization creates the presumption that the verities which are proclaimed are endowed with a validity which is universal and unchanging, even if in mathematics the claim is much more pronounced and paramount than in myths. The potency of our mathematics today derives from the fact that its symbolization is cognitively logical and, what is decisive, operationally active and fertile; myths however, from our retrospect, were always backward-directed, and their symbolizations were always remi-

[3] The large work, in 3 volumes, of Ernst Cassirer, *The Philosophy of Symbolic Forms* (1953–1957), whatever its merits, is not very helpful in this respect. Even the essay of A. N. Whitehead, *Symbolism, Its Meaning and Effect* (Cambridge University Press, 1928), has no bearing on mathematics, which is the seat of symbolism par excellence.

niscingly anthropological and operationally inert. But to Plato this disparity in the "values" of the symbolizations was, or would have been, much less accentuated than it is for us today, because in Greek mathematics, as altogether in Greek rational thinking, whatever its grandeur, *overt* symbolization was very little advanced, and its operational role was barely perceived, reflectively or pragmatically.

Outwardly, and especially to the layman, the home ground of mathematical symbolism is the so-called algebra. In depth, however, the operational efficacy of symbolization pervades all divisions and strata of mathematics equally. And present-day efforts to "modernize" mathematics in pre-college schools—be these efforts prudent or not—are nothing other than efforts to spread the acute awareness of the role of symbolization in mathematics—not only in its conceptual aspects but also in its operational consequences—into the earliest feasible stages of the schooling process.

In Greek mathematics, whatever its originality and reputation, symbolization, as an effective propellant, did not advance beyond a first stage, namely, beyond the process of *idealization,* which is a process of abstraction from "direct" actuality, be it extra-mathematical, not-yet-mathematical, or other actuality. However—as we will further elaborate in the last section of this chapter—full-scale symbolization is much more than mere idealization. It involves, in particular, untrammeled escalation of abstraction, that is, abstraction from abstraction, abstraction from abstraction from abstraction, and so forth; and, all-importantly, the general abstract objects thus arising, if viewed as instances of symbols, must be eligible for the exercise of certain productive manipulations and operations, if they are to be mathematically meaningful. From this specific approach, Plato, in his doctrine of Ideas or Forms, never advanced beyond mere idealizations, and the possibility of a hierarchical escalation is only faintly discernible, if at all. In fact, in a passage of the dialogue

Phaedo, Plato even seems to frown on the presumption that his (Platonic) Ideas of objects of mathematics should be such as to provide for mathematical operations too.[4] To put it emphatically, if by some reversal of history one of our contemporary "Parents' Guides for the New Math" had come to rest on Plato's desk, the course of Western "Idealist" philosophy might have run altogether differently.

For my part I find that the absence of a proper perception of symbolization is also acutely felt in Plato's frequent *aperçus* on the relation between poetry and truth. The relation is recondite, and, admittedly, Plato's insights are penetrating; but Plato is not at his best when, ostensibly, he castigates poetry for not being as "real" and "true" as is (Plato's "Idea" of) Reality Itself or Truth Itself.[5] In this area of endeavor his disciple Aristotle was much more circumspect and much more successful. His *Poetics* is so masterful because it does not raise the question of the relation between poetry and truth, and only tries to decide what makes poetry effective. Aristotle even omits in this book his customary historical and dialectical introduction that might lead him into altercations with predecessors. And poets and literary critics have been delighted with this book, which is always constructive and never abrasive, ever since it began to be known through printing.

Poetry is also profoundly involved with symbolization,

---

[4] It is within a larger passage, the passage 99 D–102 A; see for instance, R. Hackforth, *Plato's Phaedo,* pp. 133–136.

For a general philosopher's account of Plato's philosophy of mathematics see, for instance, Anders Wedberg, *Plato's Philosophy of Mathematics* (Stockholm, 1955). See also relevant passages in E. W. Beth, *The Foundation of Mathematics; a Study in the Philosophy of Science* (Amsterdam, 1959).

[5] Especially in Book (II and) X of the *Republic*. On the face of it, Plato deprecates poetry and visual arts. Platonists have explanations for this; see, for instance, Paul Friedländer, *Plato, An Introduction* (Harper Torchbooks, 1964), pp. 119–24, with supplementary references in note 20 on pp. 358–59.

in its distinctiveness perhaps as much as mathematics; but there seems to be a major difference between mathematics and poetry nonetheless. Since the beginning of history, rationality has been urging that science and other knowledge be articulated through mathematics; but Aristotle maintains in his *Poetics,* Chapter 1, sentence 11, that genuine poetry is not the proper medium for a discourse on "prosaic" science. And Aristotle is apparently right; although certain poems in natural philosophy, as for instance the magnificent *De Rerum Natura* of Lucretius, and even the gnarly poem of Parmenides are disturbing exceptions.

3. PRE-GREEK MATHEMATICS

In subject matter, certain parts of Euclid's *Elements,* which is the world's imperishable "primer" in mathematics, are to be found in Pre-Greek mathematics, which is the name commonly given to the mathematics of Egypt and Babylon. It is likely that the Greeks acquired these parts of Euclid by adoption, and, in fact, the Greeks themselves acknowledged, almost eagerly, that their mathematics had received its first impulses from Egypt and Babylon. Their tradition, as we have it, may have nevertheless understated the extent of the subject matter thus acquired; but even if it has, it is a fact that the Greeks endowed their mathematics, whether acquired or created, with an intellectuality of its own, and they themselves were aware of this.

Babylonian mathematics was rather advanced, even more so than Egyptian. But it is difficult to ascertain with our customary modes of rational discriminations—which are derivative from Greek thought processes—to what extent Pre-Greek mathematics arrived at a stage of being cultivated for its own sake, in a theoretical separateness, and yet, perhaps, within a general setting of intellectuality. It would be even more interesting to ascertain such matters for Mesopotamia than for Egypt, and yet in the case of

Mesopotamia the difficulties are even greater. By documentation, Pre-Greek mathematics was, preponderantly, mathematics for a purpose and enmeshed with applications. Preponderantly, it was mathematics needed in astronomy; or mathematics which occurs in computation of areas for civil purposes of taxation or inheritance, or of areas and volumes for military purposes of construction of moats and breastworks; or it was mathematics needed for international commerce, or for computation of supplies for Royal Households, temples, or armies in the field.

It is difficult to make out how Pre-Greek mathematics was taught and internally transmitted, especially to what extent the rules of mathematical procedure were inculcated by presentation of examples or paradigms and to what extent they were transmitted by some kind of "motivations," "reasonings," or elucidating comments of a general nature. The Greeks introduced names, expressive ones, for various areas of mathematics, and of other pursuits, and, above all, they created conceptions of "mathematics," "physics," "astronomy," "poetry," "rhetoric," "knowledge," "science," etc. Egyptians and Babylonians seemingly did not have such conceptions, and only very few names; but it is difficult to see, from our mode of thinking, how they could have advanced as far as they did without having had some kind of intellectual constructs which corresponded to such conceptions.[6]

From lengthy passages in Plato one can form an impression, even a vivid one, of how it sounded when Athenians of Plato's time were discussing the nature of mathematics and the peculiar difficulties of learning it; and from other indirect documentation one might even hazard to project such impressions backwards into the 5th century

[6] For an ascertainment which is very sympathetic to Pre-Greek Mathematics see Kurt Vogel, *Vorgriechische Mathematik:* I, *Vorgeschichte und Ägypten;* II, *Die Mathematik der Babylonier* (Hannover, 1959).

B.C. There seem to be no documents of this kind for Babylonian mathematics, and for Egyptian mathematics there is barely one such document, which is incorporated in papyrus Anastasi I.[7] It is a literary composition in which one royal scribe upbraids another for his incompetence in the discharge of various high offices with which he is entrusted; he especially taunts him with his inability to submit a mathematically correct estimate in certain engineering necessities which an army under his command is in need of. The mathematical passage proper has been thoroughly examined with regard to its mathematical technicalities,[8] but there has not been an examination with regard to "philosophy" of mathematics, and perhaps it would not be feasible to attempt one.

It must not be taken for granted that Pre-Greek mathematics ceases to play a part with the advent of Greek mathematics in the 6th and 5th centuries B.C., or at least with its flowering in the 3rd century B.C. External impressions suggest that certain "un-Greek" features in late Greek mathematics are due to its merger with late Babylonian mathematics, and, more intriguingly, that medieval Arabic algebra—which, chronologically, interposed itself between Greek and modern mathematics—was to an extent a descendant of Hellenic and Babylonian mathematics, both. But these suggestions are very difficult to substantiate.

4. WHAT IS IN A NAME?

The Greeks introduced names, conceptions, and self-reflections into mathematics, and very early they began to speculate on how mathematics had come into being. Their speculations were no more than a few jottings-down, but they nevertheless virtually preempted this field of specula-

---

[7] A. Erman, *The Literature of Ancient Egypt* (1927), pp. 214–234; especially the mathematical passage on pp. 223–224.

[8] O. Neugebauer, *The Exact Sciences of Antiquity,* 2nd edn. (1957), p. 79.

tion; also casual remarks of the Greeks grew into ponderous theses in the 19th century, and have become tired clichés in the 20th century.

In extant writings, the first to speculate was Herodotus (480–425 B.C.). He spoke only of *geometria,* and he was alive to its literal meaning of "land measurement"; a general notion of mathematics may not have been familiar to him. Being an anthropologist, and a sociologically oriented historian, he suggested (*Histories,* Book II, Chapter 109) that geometry came to Hellas from Egypt, where it had come into being because yearly inundations demanded that the size of landed property be frequently resurveyed for tax purposes. In the same breath he adds that the Greeks learned the use of sundials and the division of day length into 12 parts from the Babylonians. This finding of Herodotus has been over-lauded. The assertion that commonplace geometry has "utilitarian" beginnings is not very profound, and certainly not above the average level of originality and excitement of the total work of Herodotus.

Plato (427–348 B.C.) is already concerned with various parts of mathematics, and in the course of the whimsical myth of Theuth and Thamus, in the dialogue *Phaedrus* (274 C, D), he says as follows: [9]

> The story is that in the region of Naucratis in Egypt there dwelt one of the old gods of the country, the god to whom the bird Ibis is sacred, his own name being Theuth. He it was that invented number and calculation, geometry and astronomy, not to speak of draughts and dice, and above all writing.

As frequently, Plato is "whimsical" because he is not certain that he is right.

Aristotle (384–322 B.C.) at last, speaks of mathematics

[9] R. Hackforth, *Plato's Phaedrus,* The Library of Liberal Arts (1951), p. 156.

in the full meaning of the concept, that is, of mathematics for mathematics' sake and in its oneness, and he says in Chapter 1 of Book I of the *Metaphysica* (981 b 24–26) that the mathematical "sciences," or "arts" (*mathematikai technai*) originated with Egyptians because their priestly class was afforded the leisure for such pursuits. It is doubtful whether Egyptian priests constituted a leisure class of this kind, or even whether Aristotle really assumed that they did; but this does not affect the wisdom of Aristotle's observation. In the context of the chapter, the reference to the Egyptians is only meant to round out the general contention that there is a knowledge for the sake of knowledge; that pure mathematics is a prime example of it; and that the rise of this knowledge is not due to the quest for "consumer goods" and luxuries. This un-naïve Aristotelian view may be opposed, but it cannot be refuted, not convincingly, that is.

In an over-all sense, the Greeks attempted to create two methodologies of "scientific" cognition. One was their Ontology, and the other was their Mathematics; Aristotle's Logic was somewhere between the two, and by Aristotle's own intent, at any rate, was meant to be auxiliary only, in a general sense. Their Ontology was characterized by the powerful Parmenidean "Ens" (*to on, to einai*), and also by the Heraclitean nuance of "Logos," whose ontological aspect revealed itself only afterwards in its Stoic and other Hellenistic interpretations. Mathematics has far outstripped ontology as an effective methodology; but the name "mathematics" itself, for whatever reason, did not turn out to be as strong and affirmative as were, and are, Ens and Logos. Nevertheless something of the originality of the Greek creation is reflected in the manner in which the name has come about and occurs, and we will make some remarks about its meaning.

The word "mathematics" is a Greek word, and, by origin, it means "something that has been learned or understood," or perhaps "acquired knowledge," and perhaps even, somewhat against grammar, "acquirable knowledge," that is, "learnable knowledge," that is, "knowledge acquirable by learning." These meanings of the word mathematics seem to agree with the meanings of cognate roots in Sanscrit; even the great lexicographer E. Littré, who was also a distinguished classicist of his day,[10] takes notice of this fact in his *Dictionnaire de la langue française* (1873) under "mathématique." The *Oxford English Dictionary* does not refer to Sanscrit. The Byzantine Greek dictionary of the 10th century A.D., which goes under the name of "Suidas," lists words relating to "physics" and some words relating to "geometry" and "arithmetic"; but it apparently does not have the word "mathematics" directly, from "medieval darkness" perhaps.

The contraction of meaning of the word mathematics from general knowledge to mathematics proper proceeded slowly; it was only completed in Aristotle (384–322 B.C.), but not yet in Plato (427–348 B.C.).[11] This contraction is remarkable not only because it was so extensive, but because there was only one other [12] contraction, in Greek, which was equally extensive, the contraction of the word "Poetics." By origin, *poetiké* means "something that has been done, manufactured, achieved," and in the case of this word the contraction of meaning is all but completed in Plato. From whatever reasons, the lexicological and intellectual problems involved in the contraction of meaning

[10] He made the many-volumed standard edition, with translation and commentary, of the body of Greek medical works which go under the name of Hippocrates (Littré had the professional training of a physician).

[11] For some reason, in the second century A.D., or earlier, both in Greek (Sextus Empiricus) and in Latin (Suetonius) the word "mathematician" acquired the connotation of "astrologer," through the length of Later Antiquity and the Middle Ages; see the Du Cange glossaries of medieval Latin and Greek.

[12] In the case of the word "philosopher" the contraction of meaning is relatively small.

have never been raised for Poetry, nor has this peculiar parallelism between poetry and mathematics ever been noted, I think. But in the case of mathematics, the problems involved did receive some attention, and we will report on some of the developments.

First of all, statements in Aristotle in accumulation suggest that the technical use of the word "mathematics," and of cognate words, originated in the thinking of Pythagoreans, but there is nothing in the documentation to suggest that there was a similar tendency in the thinking of natural philosophers from Ionia.[13] Also, among Ionians, only Thales was credited with achievements in "pure" mathematics; and except for a brief mention in Diogenes Laertius, this accreditation comes from a source which is very late and directly mathematical, namely from the Proclus commentary of Euclid; but it does not come from Aristotle, even though he knows Thales as a "natural philosopher," or, earlier, from Herodotus, even though he knows Thales as an "activist" in politics and military tactics, who can even foretell an eclipse.[14] This may help to explain why there is relatively little made of Ionians in Plato; [15] Anaximander and Anaximenes are never mentioned; Xenophanes only once, even if meaningfully; and Thales is somebody about whom to spread anecdotes. Even Heraclitus, much as Plato may have been intrigued by his *bon mots* that "everything is in flux" and that "one cannot step into the same river twice," receives none of the reverence which Plato bestows on Parmenides, whose ontology was, methodologically, a much more serious rival

[13] For instance, B. Snell, *Die Ausdrücke für den Begriff des Wissens in der vorplatonischen Philosophie* (Berlin, 1924), pp. 72ff.

[14] G. S. Kirk and J. E. Raven, *The Presocratic Philosophers* (Cambridge University Press, 1957), pp. 75–76.

[15] E. Zeller, "Plato's Mitteilungen über frühere und gleichzeitige Philosophen," *Archiv für Geschichte der Philosophie*, Vol. 5 1892), pp. 165–89. Also reproduced in Zeller's *Collected Essays*.

to Pythagorean mathematics than was Heraclitean flux.

It appears that to some Pythagorean circles mathematics was a "way of life." In fact, from chance testimony in the Latin author Gellius [16] of the 2nd century, and of the Greek philosophers Porphyry and Iamblichus [17] of the 3rd and 4th centuries respectively, all A.D., it appears that certain Pythagoreans had a "General Graduate Program for Adults," in which there were enrolled regular members and incidental ones. The incidental members were called "auditors" (*akousmatikoi*); the regular members were called "mathematicians" (*mathematikoi*), as a general class, and not because they "specialized" in mathematics. This Pythagorean inspiration has been living forth ever since; thus, Archimedes was only and exclusively a mathematician, even to those who were transported by the uncanniness of his technical contrivances; Newton was, by academic position, a mathematician, although he was half physicist; to the general public and to journalists Albert Einstein was, preferably, a mathematician, although he was all physicist; when Roger Bacon (1214–1294) was defying his century by advocating a kind of "realistic" approach to science he was putting science into a large frame of mathematics, although, as it happens, his comprehension of mathematics in depth was very circumscribed; when Descartes, as a young man, resolved to face the future by breaking with the past he first of all shaped the name and conception of *mathesis universalis;* [18] Leibniz then took this very same conception and made it into an anticipation of the "symbolic" logic later to come; [19] and the symbolic logic of the 20th century has a hankering for being called

[16] Aulus Gellius, *The Attic Nights,* Book I, Chap. 9.
[17] For the passages from Porphyry and Iamblichus, with translation into Italian, and commentary, see M. T. Cardini, *Pitagorici, testimonianze e frammenti* (Florence, 1958), I, 84–87.
[18] In Rule IV among his *Rules of the Direction of the Mind.*
[19] See, for instance, *Leibniz Selections,* edited by Philip P. Wiener (New York, 1951), pp. xxvi–xxxii and 1–46.

"mathematical" logic, although it verily and truly is "symbolic," much more nakedly so than mathematics.

In the 18th century, Montucla, the author of the pioneering history of mathematics, says [20] that he has heard two explanations for the fact that the Greeks gave to mathematics the name of "general knowledge." One explanation has it that mathematics is intrinsically superior to other areas of knowledge; the other, that, as a topic of general instruction, it was organized first, before there was a rhetoric, dialectics, grammar, ethics, etc. Montucla adopts the second explanation, or, rather, he rejects the first on the ground that he finds no corroboration for it in the general introduction of Proclus (5th century A.D.) to his commentary on the first book of Euclid [21] or in any other statement from antiquity. Nevertheless, etymologists in the 19th century were inclining to the first explanation, but classicists in the 20th century lean again to the second.[22] For my part, I find that the two explanations are quite compatible with each other and I am opting for a fusion of the two. Mathematics started very early, and its superiority is *sui generis*.

5. UNIVERSALITY OF MATHEMATICS

When Leibniz undertook to "formalize" by "logification" Descartes' *mathesis universalis,* he half-expected that it would soon be generally adopted. In this expectation he was disappointed. The efforts of Leibniz were noted by others, even approvingly, but in the 18th century his calls for action were not heeded. But the 18th century made an

[20] J. F. Montucla, *Histoire des mathématiques,* 2nd edn., 4 vols. (Paris, 1799–1802; reprinted 1960), I, 1–3.

[21] As against this I would like to observe that Proclus, as a neo-Platonist, was simply bound to subscribe to the Platonic view, which is expressly imputed to Plato by Aristotle in his *Metaphysica* (mainly in Book I, Chap. 4, 937 b 15–19), that objects of mathematics, qua "ideas," are intermediate between "sensibles" and True Forms, and not True Forms themselves.

[22] See n. 13.

entirely different contribution to a re-universalization of mathematics when Immanuel Kant performed an apriorization of geometry and arithmetic in the "Transcendental Aesthetic" of his *Critique of Pure Reason* (1781, 1787), whatever Kant's own philosophical intent may have been. Kant's work would have retained much more pertinence for the 20th century if it had not limited its apriorization to the very circumscribed topics of Euclidean figures and numbers but had penetrated through to the cognitive apriorisms of algebraization and symbolization of mathematics, which, internally though not yet manifestly, were already fully operative in the mathematics of the day, namely, in the analytical setting of functions and in the coordinate setting of geometry. In the 17th century such "modernisms" were yet out of sight. Even Galileo in his famous dictum that "mathematics is the language of science" says only that the book of nature "is written in mathematical language, and the characters are triangles, circles, and other geometric figures." [23] However, in the second half of the 18th century, the modernisms might have been much more discernible, and if Kant had been inspired to orient his basic apriorizations by the mathematics of his times, the course of philosophy of knowledge would have run differently.[24]

In the 19th century a new wave of universalization of mathematics was triggered by Georg Cantor's creation of the theory of sets, which is also the staging area of the "New Math" in our present-day schools. The creation of set theory was more meaningful than the synchronous creation of symbolic logic, because it furthered the universalization much more by expansion from within than by

[23] G. Galilei, *Il Saggiatore* (*Opere*, VI, 232): "è scritto in lingua matematica, e i caratteri son triangoli, cerchi, ed altre figure geometriche."

[24] In his early work: *Gedanken von der wahren Schätzung der lebendigen Kräfte* (1747), Kant is more aware of the mathematics of his times than in the celebrated *Critique*.

accretion from without. In the 20th century a very manifest and intentional trend to universalize mathematics has become discernible, in the second half of the century at any rate. It is much more than an attempt to resume where Leibniz left off; but it is not easy to say how the trend has come about or to speculate on whether it will continue, other than in easy slogans and extrapolations perhaps. In large part, and outwardly, this trend manifests itself in a widespread disposition and desire to "apply" mathematics wherever and whenever possible, and the mere fact that mathematics is used seems to give some kind of satisfaction. But it is also a fact that only a fraction of contemporary high-level research activity in mathematics, whatever its eventual impact, is motivated and instigated directly by applications. Much of newly created mathematics has, at the time of its creation at any rate, no overt bearing on technology, or on applied science, or even directly on theoretical science, and yet it somehow seems inconceivable that such mathematics should not have some purpose and significance within the texture of our total knowledge. Also, all this makes even more acute the question, the perennial one, which is asked sometimes shyly and sometimes challengingly, why it is that mathematics is so peculiarly "abstract," in a manner which is discouraging to many who try to learn it, and whether mathematics cannot be made more "intuitive," and more accessible to larger audiences, without forfeiting its substance and its mission, whatever these might be. Questions like these are of the kind to which answers can only be attempted, but must not be expected.

6. SOME CONTRASTS BETWEEN GREEK AND PRESENTDAY MATHEMATICS

The "classical" Greek mathematics is known from, and culminated in, the works of Euclid, Archimedes, and Apollonius, and its structure and importance are confirmed in the majestic astronomical treatise of Ptolemy. Somewhat different in tone, perhaps only because it is intended to be "applied" mathematics, is the work of the "theoretical engineer" Heron. The only major extant work which is distinct in texture is the number-theoretic work of Diophantus, and it is a sad fact that nothing is known about its evolutionary relation to the geometrically controlled work of the others.

All these works played a great role in the outburst of science and new mathematics in the 16th and 17th centuries, and in the very formative 17th century they were studied and exploited as never before. Yet it is a hard fact that within Hellenistic civilization itself mathematical activity eventually ceased entirely, and so-called exact science and technology, in a proper and general sense, did not come into being at all. There are various "cultural" and "accidental" reasons which historians adduce for this singular and definitive failure of Greek civilization, but somehow none is really convincing. Greek astronomy was mathematical from the outset, and its mathematization made it highly successful; the Greeks originated a doctrine of optics, and its geometrization endowed it with possibilities of development; and, already beginning with earlier Pythagoreans, there was some degree of awareness abroad that a mathematization of concepts and relations would be an effective approach to the comprehension of nature. And yet, from whatever reasons, the Greeks never penetrated to the insight that an effective manner of pursuing mechanics, physics, and other science is to articulate qualitative attributes by quantitative magnitudes, and then to represent connecting rules and laws by mathematical formulas and relations in order that the application of mathematical procedures to the formulas and relations may lead to further explications and developments.

The Greeks never arrived at such an insight; in spite of striking first steps in Archimedes they never developed a "theoretical" science in general, and eventually they even lost control over their mathematics, which they set out creating magnificently. Perhaps the sharpest illustration of the Greek "failure" can be obtained by a contemplation of the collection of works which goes under the name of Heron of Alexandria—and for our purposes we may assume, what some historians doubt, that there was one such person who was the author of the bulk of it. Now, the author of this collection was an eminent mathematical engineer, and it is remarkable that, like other great scientific engineers after him, he apparently wrote his own textbooks in mathematics, both on the elementary and on an advanced level. He was a great inventor of "trap doors," "magical devices," and general "wonders," but behind all this there was no proper scientific and technological urge, in our sense, that would drive this inventiveness into channels of utilization, industrialization, or other exploration, and Heron's great talent virtually came to naught. Somehow, there was no determined scientific or technological "motivation" within the totality of knowledge and effort, and this absence of a proper scientific motivation was so fateful that eventually even Greek mathematics withered away.

It appears that present-day mathematics, and mathematically oriented science with it, grows and thrives in spite of, and perhaps even because of, a certain fruitful self-contradictoriness in its pattern of unfolding. There is a quest, a dominant one, for a mathematics-for-its-own-sake, which is motivated internally and almost self-sufficiently. Parallel to this there is a quest, a strong one, for a mathematics-for-purposes, which is motivated mostly externally. Also, for the most part there is no forcible conjunction of the two quests, or even of the responses to these different quests. Yet the total outcome is very gratifying, even though there is, on the whole, no forcible effort towards reconciling the divergences of the quests by direct and embraceful confrontations.

The Greeks had an uneasy premonition of all this. But they were the original and proud discoverers of the meaning of "logicality," in and out of mathematics, and it would have been intolerable to them to countenance any kind of "self-contradictoriness" in connection with mathematics, not even in an external genetic aspect of its growth. When posted on their frontiers of knowledge, the Greeks were rigid observers of all rules of discipline and of what they deemed to be "logical" conduct, in spite of the fact, or because of the fact, that all these rules had been drawn up by themselves.

7. EUCLID'S ELEMENTS

Since Greek mathematics as such did not have the genetic stamina to survive, it becomes doubly fascinating to contemplate the extraordinary career, which has never entirely ceased, of Euclid's unique treatise in 13 "books." He was not one of history's greatest mathematicians, and even antiquity did not misconceive Euclid's own level of research originality. Professional mathematicians admired his work, but nobody would have likened him to Archimedes, say. Antiquity already knew that Archimedes was one of the greatest men of scientific knowledge ever, and his name was familiar not only to circumscribed circles but to the "educated public" at large.[25] Cicero (106–43 B.C.) and Plutarch (A.D. 46–120) offer extensive testimonials to the greatness of Archimedes. Furthermore there are "notices" about Archimedes, of varying length, in the works —we list them more or less chronologically—of the historian Polybius (202–120 B.C.), the geographer Strabo

---

[25] To this and the following compare the Introduction in the book of T. L. Heath, *The Works of Archimedes* (Cambridge University Press, 1897).

(69 B.C.–A.D. 19), the historian Livy (59 B.C.–A.D. 17), the architect Vitruvius (50–26 B.C.), the historian Diodorus (40 B.C.), the philosopher of religion Philo (A.D. 39), the astronomer Ptolemy (A.D. 150), the anatomist Galen (A.D. 129–199), the philosopher of scepticism Sextus Empiricus (A.D. 190), the columnist-gourmet Athenaeus (A.D. 200), the Church Father Lactantius (born A.D. 250), the mathematician Pappus (A.D. 320), the historian Ammianus Marcellinus (A.D. 390), the Latin writer Macrobius (A.D. 400), the poet Claudian (A.D. 400), the writer Martianus Capella (early 5th century A.D.), the neo-Platonist and Euclid commentator Proclus (A.D. 411–485), and finally, very late, the promiscuous Byzantine writer Tzetzes (A.D. 1110–1180).

There is nothing like this in the case of Euclid. In comparison to Archimedes, Euclid was virtually unknown. Nothing is known about him as a person. There are only two little anecdotes about him, and they are uninformative and banal. Their "philosophical" content is trivial when held against pronouncements on mathematics by Plato and Aristotle before him, and even by Herodotus long before him. In one anecdote, Euclid chides a student of his, by a gesture and not by a rational explanation, for wanting to know wherefore he should study a proposition which was apparently assigned to him as homework. In the second anecdote, Euclid states before the King of Egypt that there is no Royal Road to mathematics; even a mighty ruler cannot command that he be taught mathematics in Ten Easy Lessons.

Yet Euclid's work has survived "intact," that is, without any manifest stigmata of textual corruptions, as few large works from classical antiquity have, although it is not a book that stirs "human interest" as do the works of Homer or Herodotus. It does not speak of human passions as do the all too few extant plays of Aeschylus, Sophocles, and Euripides, and it does not challenge the intellect with the

[34]

immediacy and directness of the works of Thucydides, Plato, and Aristotle, and also of the fragments of Parmenides. It is not even a book that has a direct scientific interest, or the same appeal to adultness, as Ptolemy's treatise on astronomy; it is not at all a book of "technological" interest as are the books of the large compilation which is attributed to somebody named Heron. Also, if examined "objectively," Euclid's work ought to have been any educationist's nightmare. The work presumes to begin from a beginning; that is, it presupposes a certain level of readiness, but makes no other prerequisites. Yet it never offers any "motivations," it has no illuminating "asides," it does not attempt to make anything "intuitive," and it avoids "applications" to a fault. It is so "humorless" in its mathematical purism that, although it is a book about "Elements," it nevertheless does not unbend long enough in its singlemindedness to make the remark, however incidentally, that if a rectangle has a base of 3 inches and a height of 4 inches then it has an area of 12 square inches. Euclid's work never mentions the name of a person; it never makes a statement about, or even an (intended) allusion to, genetic developments of mathematics; it makes no cross references, except once, the exception being in proposition 2 of Book 13, where the text refers to, and repeats the content of, the "first theorem of the tenth book," which, as it happens, is Euclid's "substitute" for the later axiom of Archimedes. Euclid has a fixed pattern for the enunciation of a proposition, and, through the whole length of 13 books, he is never tempted to deviate from it. In short, it is almost impossible to refute an assertion that the *Elements* is the work of an unsufferable pedant and martinet.

On the face of it, in comparison to the *Elements,* the works of Archimedes, Apollonius and Ptolemy are "pulsating with life." They address themselves not to students but to regular professionals, have a considerable research level to begin with, are mathematically more difficult to approach, and their style is sometimes over-solemn, but their manner of presentation, although very controlled, is by no means rigid. There are motivations, explanations, justifications. There are historical passages, with names of past scholars, and those in Ptolemy's *Almagest* are of immeasurable value. The *Sand-Reckoner* of Archimedes is addressed throughout to King Gelon. The epistolary preambles in Archimedes and Apollonius have personal touches; they are informative, and sometimes uninhibited or even chatty.

In an apparent effort to emulate, or comply with, certain Aristotelian philosophical doctrines of his day, Euclid attempted, at the head of his Book 1, to separate certain primary data into "postulates" and "common notions," but generations of commentators have not been able to rationalize this separation satisfactorily. Archimedes, however, ignored these would-be philosophical distinctions as if they were schoolmasterly pedantries, and he uses such related terms indiscriminately, as the fancy strikes him.[26]

Euclid's work became one of the all-time best sellers. According to "objective" Pestalozzi criteria, it should have been spurned by students and "progressive" teachers in every generation. But it nevertheless survived intact all the turmoils, ravages, and illiteracies of the dissolving Roman Empire, of the early Dark Ages, of the Crusades, and of the plagues and famines of the later Middle Ages. And, since printing began, Euclid has been printed in as many editions, and in as many languages, as perhaps no other book outside the Bible.[27] Euclid lived through the Scientific Revolution of the 17th century unscathed, and even the

[26] See K. von Fritz, "Die ARCHAI in der griechischen mathematik," *Archiv für Begriffsgeschichte,* Vol. 1 (1955), pp. 13–103.

[27] T. L. Heath, *A History of Greek Mathematics,* 2 vols., I, 354ff. See also T. L. Heath, *The Thirteen Books of Euclid's Elements,* 3 vols.; especially I, 1–6.

discovery of non-Euclidean geometries in the first half of the 19th century did not tarnish the "image" of Euclid, or of his *Elements*. American towns and cities continued to name streets and avenues after Euclid, much more so than after Plato or Aristotle. And Abraham Lincoln, in his campaign biography of 1860, written by himself and published under the name of John L. Scripps of the Chicago Press and Tribune, ventured to assert about himself that "he studied and nearly mastered the six books of Euclid since he was a member of Congress." [28] Lincoln's assertion that he had "nearly mastered" these books was one of the boldest and blandest campaign statements in the annals of American presidential elections, and folkloristic embellishments of this assertion were even less restrained.[29] It is worth reflecting on the fact that in the America of 1860 a consummate grass-roots politician of the then Mid-Western Frontier should have thought that adding to a mixture of log cabin and railsplitting a six books' worth of Euclid would make the mixture more palatable to an electorate across the country.

Even the 20th century has its problems with Euclid. The present-day reformers, all over the world, of mathematical programs in primary and secondary schools have not really been able to decide with unrevocable finality what to do about the unreformable Euclid in the midst of it all.

8. MATHEMATICS ON THE RISE

Mathematics, although shielded by a texture of unchangeability, was likewise caught in the Great Scientific Revolution, which spanned the 16th, 17th, and 18th centuries,

---

[28] *The Collected Works of Abraham Lincoln*, The Abraham Lincoln Association, Springfield, Illinois (Rutgers University Press, 1953), IV, 62.

[29] See Herndon's *Life of Lincoln* (The World Publishing Company, 1949); Carl Sandburg, *Abraham Lincoln, The Prairie Years* (Harcourt, Brace, 1926), I, 423–424; Emanuel Hertz, *Lincoln Talks* (Viking, 1936), p. 18.

and centered in the 17th century. Mathematics then began to go apace, as if anew and from novel sources; it revealed novel springs of operational efficacy; and it assumed a novel stance of indispensability for science and knowledge at large. Yet, in spite of this, the working techniques of Greek mathematics were studied and scrutinized as never before and explored and exploited to the limit of their yield. An everlasting monument to Archimedes was erected when Isaac Newton, his greatest rival within the genetic line from Pythagoras to Planck, cast his forward-thrusting *Principia* (1686) in the mold of much-tested though backward-directed Archimedism. Even the semi-folkloristic mathematics of Diophantus became what it is today through the intercession of the lay-mathematician Pierre Fermat in the 17th century. Even more significant for the "problematics" of the role and nature of mathematics is the fact that for the mathematics of the era of the Scientific Revolution it would have been very difficult to say at the time what of it was mathematics to a purpose and what was mathematics for its own sake. Most of it was, in fact, mathematics to a purpose, immediately or eventually. But we contend that contemporaries would not have been able to assess degrees of purposefulness, even if they could have been made to comprehend the nature, wisdom, and cogency of such assessments. We will adduce three arguments in support of this contention, one from algebra, another from geometry, and a third from mechanics.

(1) In mathematics, the 16th century was the century of the rise of so-called algebra, whatever that be. Not only was this algebra a characteristic of the century, but a certain feature of it, namely the "symbolization" inherent to it, became a profoundly distinguishing mark of all mathematics to follow. This distinction is so pronounced that one sometimes encounters the (superficial) judgment that if only the Greeks had happened to "discover" this

algebra, the course of history of mathematics would have run differently. At any rate, this feature of algebra has become an attribute of the essence of mathematics, of its foundations, and of its nature of abstractness on the uppermost level of "ideation" *à la* Plato. Yet, if one tries to uncover where this algebra came from, what its motivations were, and, above all, why it grew, then apparently the only explanation ever proposed is this: that, as a twin phenomenon to "arithmetic," it was brought into being, preponderantly, by the Rise of Commerce in the late Middle Ages; so that the challenges to which the rise of algebra was the response were predominantly the very unlofty and utilitarian demands of counting houses of bankers and merchants in Lombardy, Northern Europe, and the Levant.[30] Even Oswald Spengler, whose views run counter to those of economic determinists of whatever sociological convictions, when attempting to demonstrate in his essay on Mathematics[31] that modern mathematics has "cultural" roots in medieval Gothic, is concerned more with mathematical functions than with Renaissance algebra. His attempted derivation is attuned much more to mathematical functions in analysis than to symbolization in algebra, although, on mathematical grounds, the rise of functions and the rise of algebra are closely linked.

(2) The case of coordinate geometry runs almost in reverse. Nowadays there is hardly anything that is more

[30] Wallace K. Ferguson, *The Renaissance* (1940); the first systematic presentation of this explanation is perhaps due to Werner Sombart; see his book *The Quintessence of Capitalism* (1915), Chap. 8, "The Art of Calculation."

The fact that Renaissance mathematics was stimulated by needs of merchants rather than by those of "industrialists" need not surprise; in the late Middle Ages of the West, the general economy was dominated not by the producer, but by the trader. See Marc Bloch, *The Feudal Society* (London, 1961), p. 71.

[31] *The Decline of the West*, Vol. I, Chap. 2, "The Meaning of Numbers."

commonplace and utilitarian than a "pictorial" graph or chart, especially when drawn with a pencil on a sheet of pre-latticed graph paper. It is incomprehensible why graphs were not introduced, for instance, by Later Greek mathematics, which obviously "fraternized" with Babylonian astronomy, in which a tabulatory approach to would-be graphs of would-be functions is clearly discernible.[32] It is "inexcusable" that graphs, apparently, began to be used only in the second half of the 18th century, as an outgrowth of the employment of geometric coordinate systems in the first half of the 18th century, first in mathematics and then, gradually, in rational mechanics, for which they became the indispensable setting. Coordinate systems, in their turn, have become the hallmark of the "new" geometry that had been inaugurated by (Fermat and) Descartes in the first half of the 17th century, so much so, that the name "Cartesian" has been firmly affixed to them, although coordinate systems in the full sense of the notion only *begin* to occur in the work of Descartes himself. Now it may be stated that there was nothing "practical" or "utilitarian" in Descartes' own conception of his geometry nor in the conception of his contemporaries or immediate successors. Descartes motivated his innovation on abstractly theoretical and even philosophical grounds and, somehow against his usual posture of progressivism, more as a reaction to past mathematics than as a forward acting departure from it. To his admiring contemporaries, the theory appeared to be so esoteric and recondite that regular commentaries to it were composed.[33] Mathematically, the most immediate purpose of Descartes was to initiate an appropriate technique for what eventually became known as "algebraic geometry." It is very

[32] See Chap. 5 of the book of O. Neugebauer, cited in n. 8.
[33] To all this compare C. B. Boyer, *History of Analytic Geometry* (1956).

remarkable that Isaac Newton, who comprehended the new geometry as deeply as anybody else, Descartes included, applied it in fact only to a specialized study in algebraic geometry.[34] In the *Principia,* and in the *Opticks,* there is no sign of an anticipation that pre-introduced coordinate systems would become the ineluctable setting in all of mechanics, a century hence, and in all of physics, afterwards.

(3) Finally, our argument from mechanics is as follows: For the 18th century and in the first decades of the 19th century the amount of mathematics which was created for, and introduced into, mechanics, theoretical and applied, was enormous, to the discomfort of many a "practical" scientist of the day, no doubt. For instance, in engineering mechanics nothing has a more "practical" sound than words like "stress," "strain," "pressure," "torque," etc. Yet it is a hard fact that the present-day systematic theory of these concepts was created in the 1820's by A. Cauchy, and that Cauchy was, by all odds, a mathematician first and foremost, and a very "pure" mathematician at that. In the 18th century, only Daniel Bernoulli was a physicist, recognizably so. But virtually all other leading architects of the various parts of mechanics of discrete and continuous media were just as recognizably mathematicians, and among the most eminent since Pythagoras too: James and John Bernoulli, d'Alembert, Euler, Lagrange, Poisson, Laplace, Gauss, Cauchy, Jacobi, Hamilton, etc. Correspondingly, most of their theorizing emanated from "pure thinking," with very circumscribed entanglement in direct planned experimentation, and not from pure thinking of "theoretical" physicists qua physicists, as were Gibbs, Boltzmann, and Planck a century later, but from pure thinking of mathematicians who, being mathematicians,

[34] *Enumeration of Lines of the Third Order,* translated by C. R. M. Talbot (London, Bohn, 1861).

were confident of being able to do mechanics as well. Yet the mechanics that has evolved *and* the mathematics that was created in the process are for the ages.

9. PREFABRICATED MATHEMATICS

Not all mathematics from the 17th and 18th centuries, even when done by mathematicians who excelled in mechanics, was originally mathematics to a purpose, and some of it has remained mathematics for its own sake until today. Then with the beginning of the 19th century, mathematics became wholly independent of mechanics and then physics, and its intellectual self-sufficiency began to be a philosophical verity. One began to encounter shadings of the eventual statement that, in the schema of our knowledge, mathematics ought to be counted not among the Natural Sciences (*Naturwissenschaften*) but among the Humanities (*Geisteswissenschaften*). Also, Galileo's commonplace dictum that "mathematics is the language of science," which even influences college curricula, is a partial corroboration of this statement, since it brackets mathematics with language. Yet, although the old intimacy between mathematics and mechanics was seemingly extenuated, another kind of relationship between mathematics and theoretical physics, in lieu of the old one, was beginning to develop. It was a rapport built more on parallelisms of pursuits rather than on identities of aims. The fruitfulness of the rapport manifests itself in the phenomenon that, from time to time, theoretical physics is able to seize upon a ready-made piece of mathematics, which has not been familiar to it before, and to turn it to instant use, as if it had been prefabricated for just the occasion. Mostly, such prefabricated mathematics is, by origin, mathematics for its own sake, and yet its yield, when it is invoked, is usually richer than that of so-called applied mathematics, that is, of mathematics which is mathematics to a purpose by intent and design.

A known illustration of all this, the first in this century, occurs in the Theory of Relativity. The Special Theory of Relativity did not pre-require mathematics that was unfamiliar to physicists; even Minkowski's placement of existing formulas from physics into a framework of four-dimensionality did not come about outside the mathematics already familiar from the theories of Maxwell, Lorentz, and Poincaré. But when Albert Einstein was pondering the formulation of the General Theory, he was lacking a mathematical technique that would lead him toward it. But then a mathematician directed his attention to recent work in tensor analysis by G. Ricci and also T. Levi-Civita.[35] As it happens, this "analysis" had arisen in a context from theoretical mechanics. However, the subject matter had been a mathematically esoteric one; physicists in general had had no provocation to become familiar with it, and it was mathematics for its own sake by general tenor. Einstein, however, found it to be a mathematical device most providentially prefabricated for his purposes, and he adopted it with gusto. This analysis then entered the very texture of the theory, which became inseparable from it. Furthermore, this novel analysis imparted to physics as a whole a certain procedural flexibility, and at the same time it directed the attention of mathematicians to a significant theory, as yet little noticed, in their own midst. There developed a (limited) collaboration between physicists and mathematicians, which, in this specialized area at any rate, has remained unbroken, and mathematicians began earnestly to cultivate the novel "differential calculus" for its own sake. The calculus steadily gained in significance, and eventually it became important to the field of Topology in pure mathematics. Furthermore, the theory of relativity, as cultivated by physicists, began to be interested in areas of topology in which the tensor calculus had domiciled itself, so that a new and even more

[35] Philip Frank, *Einstein, His Life and Times* (1953), p. 103.

sophisticated contact between relativity and differential geometry has come about.[36]

In quantum physics, which is the core of science, it happened also that a setting was fashioned out of pre-fabricated mathematics, not immediately but rather soon. The original disparate versions of W. Heisenberg and of E. Schrödinger were soon merged by Schrödinger into one, and the union of the two was mathematically consummated, undivorceably and fruitfully, in the precincts of so-called Hilbert space. The notions, in quantum physics, of "physical state," and of "observable," are vague, disquieting, and never-to-be-assimilated. But their mathematical counterparts, namely the notions of "vector in Hilbert space" and of "linear operator in Hilbert space" are determined, reassuring, and assimilable. D. Hilbert had evolved his theory about 20 years previously. It was a magnificent theory, but mathematicians did not quite know whither to carry it further. Since entering physics, however, the theory of operators has been studied unremittingly by mathematicians and physicists alike. Also, the preoccupation with it has made the concept of an operator from a tool in physics to a reality in nature, and it has raised the mathematization of physics to new levels. There is hardly a purely mathematical statement on operators in Hilbert space which some physicist would not, by a transliteration, interpret as an event, or as a property of an event in nature.

Furthermore, quantum physics appropriated for itself large parts of group theory, so much so that physcists began to compose their own mathematical monographs about it, or at least to give their own accounts of what is relevant to physics.[37]

[36] J. A. Wheeler, *Geometrodynamics,* Italian Physical Society (Academic Press, 1962).

[37] One of the first of this kind was the book of E. P. Wigner, 1931, whose translation in 1959 is entitled, *Group Theory and Its Application to the Quantum Mechanics of Atomic Spectra.*

A very different case of intervention of prefabricated mathematics occurred in the theory of shock waves. A knowledge of the theory became urgent with the advent of atomic fission in the 1940's, and it was found that the best extant study about this subject was a full-length book that had been written in 1903 by the eminent mathematician J. Hadamard.[38] The book in its turn was based on a much earlier memoir by the great 19th-century mathematician B. Riemann [39] and subsequent work by others. In 1903 Hadamard was at the height of his career, and he wrote a book which became topical only 40 years afterward.

Finally we may point out that contemporary studies in network and information theory, mechanical and human, in their beginnings had to fall back on work of George Boole (1815–1864) from a century ago. The title of Boole's work, *Laws of Thought,* suggests that he did not underestimate its level, but there is no intimation that the work of Boole had been done with such specific applications in prospect as the 20th century has found for it.

10. RELATION BETWEEN MATHEMATICS AND SCIENCE

It is not easy to explain how it comes about that every so often it is possible to apply, directly and potently, a mathematical development of yesterday to an actuality in science of today; the problem of explaining this possibility has been, in one formulation or another, a central problem in the Theory of Knowledge since Plato. Before Plato, the problem already occurs in the poem of Parmenides. But Parmenides was an ontologist, the first and greatest ever, and the problem appears, without any reference to mathematics, as an inconsistency between the immutability of the (Parmenidean) Being in the thinking about it and its obvious variability in the sense perceptions of it. Zeno of

[38] J. Hadamard, *Leçons sur la propagation des ondes* (1903).
[39] B. Riemann, "Über die Fortpflanzung ebener Luftwellen von endlicher Schwingungsweite," *Gesammelte Werke,* p. 145.

Elea, the renowned disciple of Parmenides, began to see the problem as one affecting mathematics. However, it appears from the main report about Zeno in Aristotle's *Physica,* and from other evidence about him, that Zeno envisaged the problem in certain "puzzling" subproblems, and not in one comprehensive tableau. The "puzzles" of Zeno are inexhaustibly fascinating to philosophy, but the "puzzliness" distracts from the fact that underlying it all is the problem of the organic relation between mathematics and science, which, as a problem, is bigger than any one of the puzzles.

To judge by Aristotle's *Metaphysica* (Book I, Chapters 5–8) Pythagoreans were the first to raise the problem, in pre-articulate versions, of the validity and significance of mathematical knowledge for other knowledge, especially for scientific knowledge. Plato inherited their problem and articulated it better; and, as can be seen from the dialogue *Meno,* at some time Plato widened the problem by merging with it the Socratic problem of establishing a social and economic doctrine by rationalization from logicality. All this contributed to the supremacy of Greek philosophy and to the perdurance of Greek philosophemes; but the problem of the relation between mathematics and science remains.

Cognitively, mathematics is not part of science, or subordinate to science as a whole. A scientist in his working mood may please himself to view mathematics as nothing other than a "handmaiden" of science; but, philosophically, this is a shallow statement. Ontologically, mathematics and science have different subject matters, and the difference cannot be bridged, unless by a deliberate identification. Any natural science deals with the "external world," whatever that be. This even applies to the psychology of the human mind, say. In a scientifically conducted psychological introspection or self-inspection, the internal object of the inspection becomes "external-

[46]

ized" by the procedure itself. Mathematics, however, deals, strictly speaking, only with objects of its own "aesthetic" perception and aprioristic emanation, and these objects are internally conceived, internally created, and inwardly structured, even though they are tangible, substantive, selfsame, and somehow communicable and shareable.

When one is concerned only with the effect of mathematics on science, the farthest one can go, cognitively, in subordinating mathematics to science is suggested by the following simile. If science is viewed as an industrial establishment, then mathematics is an associated power plant which feeds a certain kind of indispensable energy into the establishment. The counterparts to mathematicians would be the designers, maintenance men, and administrators of the power plant. Of these, the majority need be interested only in the requirements of the power plant itself, and solely a minority need be aware of the actual workings of the establishment itself, let alone be expert in its activities.

This simile is unfavorable to mathematics in that it does not account for the great power of creativity which resides within its compass. We wish to say, however, that there also is danger in over-glorifying mathematics and attempting to subordinate all of science to it, naïvely, hastily, and forcibly. Pythagoreans, and even Plato, after having discovered, or created, the metaphysical status of mathematics, were prone to endow mathematics with a somewhat lop-sided priority in the realm of cognition, and I think that genuine scientists like Archimedes and Ptolemy were sceptical of such exaggerations and kept out of such philosophers' ways. In some respects it is almost safer for mathematics to remain the handmaiden of science than to be forcibly enthroned as the dictatorial mistress of it; yet, in our very presence, that is, in the second half of the 20th century, there seem to be trends developing to bring about just that.

[47]

11. THE ABSTRACTNESS OF MATHEMATICS

We are returning, in amplifications, to a topic already broached in section 2 on "Mathematics and Myths."

There is a statement abroad that mathematics is abstract, and the statement is frequently made with emotion, either approvingly or disapprovingly, or critically with the implication that the abstractness is carried too far and ought to be kept within limits. Apart from emotional value judgments, if the statement itself is right, then objectively it can only mean that in mathematics abstractness has a quality of its own which is peculiar to it. We will make a brief examination whether such is indeed the case.

In general, to think "abstractly" means to exercise a certain mental faculty of intellectual selectivity, to bring about suitable selections, and to concentrate one's intellectual attention on what has been selected. Now any rational thinking has to be selective and concentrative, and thus, in a sense, all rational thinking is abstract thinking. The cursory reading of an ordinary telephone directory, or of a very plain listing of historical or geographical names and facts, is perhaps not yet an "abstract" process, not discernibly so at any rate, but the reading of any other book is. A cookbook is abstract, and so are the Yellow Pages in a telephone directory. Any academic textbook, on whatever level and in whatever subject, must be abstract in a more than perfunctory sense or it is useless. This general value of abstraction does not yet single out mathematics to any extent, and is obviously not what is meant when it is said that mathematics is too "abstract."

One might say that what makes mathematical abstraction different is the formation of many difficult conceptions whose definition and scope must be meticulously adhered to. But this is far from being characteristic of mathematics. It also applies very much to jurisprudence, and sharply so; and a good case can be made out for the view that in

Greece the sharply logical thinking which, to us, began with Socrates had actually originated in forensic rhetoric.[40] Also, for instance, contemporary literary criticism has become severely conceptualized, so much so that a non-professional reader frequently may find it easier to understand a literary work directly, even a recondite one, than to grasp a piece of literary criticism about it. Furthermore, in sociology, for instance, conceptualizations, and even incomprehensibilities, seem to grow more profusely than anywhere else.

At this stage one may be tempted to say that if stringent conceptualization is not yet characteristic of mathematical thinking, then its linking and pairing with symbolization is, and we mean a bona fide, outwardly recognizable symbolization, in print, with which we are all familiar, and not an implicit and internal symbolization of the kind which we have imputed to myths above. Now, the use of outward symbols has certainly a powerful place in mathematics; but, as one can quickly see, if the manifest use of symbols is to be made into a significant part of the characterization of mathematical thinking, considerable clarifications have to be proffered. Chemical theories of valency and bond are steeped in a highly conventionalized symbolism from which they could not be separated. But it is not a symbolism which is interlocked with mathematization; in fact, in fully theoretical works on quantum chemistry the "typical" symbolism of chemistry is displayed less emphatically than in regular manuals of chemistry.

Furthermore, as is well known, the classical Greek mathematics of Euclid, Archimedes, Apollonius, and others does not operate with symbols at all, not even in "disguise," as did Cardano's in the 16th century, which virtually had formulas of our kind, but expressed them in running sentences and not through formulaic symbols. This

[40] See Ernst Kapp, *Greek Foundations of Traditional Logic* (New York, 1942); also Cicero, *Brutus*, I, 1–36.

is even viewed as a defect of Greek mathematics, and as one which causally brought about the decay of Greek mathematics in the end. Yet it would not be proper to say that, in general, the Greeks did not have an affinity to thinking in symbols, mathematical or related ones. This is simply not true; in a certain qualified sense the Greeks even had a marked affinity to thinking in symbols. For instance, by all evidence it was the Greeks who were the first to use the letters of the alphabet as "symbols" for numerals. The alphabet itself was not of their own creation; this they took from the Phoenicians. However, the Phoenicians themselves never used their alphabet for numerical purposes but had separate signs for numbers, whereas the Greek installation of the alphabetical system of numerals is presumed to reach back into the 8th century B.C.[41]

Also, Aristotle in his *Prior Analytics* employs letters as symbols for logical data, and he is not in the least self-conscious about it; neither Aristotle himself nor anybody else in antiquity saw anything unusual or original in this. Some modern commentators, however, have been much impressed by this. Paul Tannery[42] notes that Aristotle uses letters as symbolic representations of objects of thought, and to Tannery this means that the Greeks were only a step away from the algebraic algorithm of Viète. He is distressed over the fact that Pythagoreans did not take the same one step as did Viète but instead abandoned themselves to sterile reveries. J. Jørgensen[43] echoes Tannery when saying that "in his treatment of propositions and syllogisms, Aristotle uses letters as symbols for the terms, thus laying the first foundation of all subsequent symbolic logic and algebra." For my part, I cannot at all agree with these two statements. It is true of course that the Greek employment of letters in lieu of numerals, propositions,

---

[41] See Heath (n. 4), pp. 32–33.
[42] *Oeuvres scientifiques*, IV, 199.
[43] *A Treatise on Formal Logic*, Vol. I (1931), p. 47.

and syllogisms falls within the general area of mathematical abstraction, and it does indeed testify to a measure of Greek capacity for it; but it is not, or not yet, the kind of abstraction which constitutes the essence of the mathematics of today. The difference between the two levels of abstraction, ancient and modern, is a very considerable one, and not much is gained by viewing it as a difference of a single cognitive step only, which the Greeks somehow failed to take, as if by accident. We repeat from section 2 above that, as viewed in retrospect, the Greeks, for all their cleverness, were not able, or not yet able, to make abstractions that were more than idealizations from immediate actuality and "external" reality; their abstractions were rarely more than one-step abstractions, and as such they remained within the purview of what is called "intuitive" in an obvious and direct sense. They did not make second abstractions from abstractions or, alternately, abstractions from intellectually conceived possibilities and potentialities, and such abstractions of higher order as were made remained rudimentary and operationally unproductive. In fact, strange as it may seem, our present-day notion of "abstract" somehow remained alien to the classic Hellenic and Hellenistic idiom of philosophy; present-day English-Greek dictionaries do not find it easy to translate the word "abstract" into classical Greek.

The Greeks clearly knew that a geometrical line is the idealization of a physical wire whose non-zero diameter is allowed, ideally, to shrink to zero. Archimedes was more or less in possession of the notion of a mass point, knowing that it is an "ideal" limit of a small ball of variable radius but fixed mass whose radius has shrunk to zero. The Greeks abstracted a notion of a "general" triangle from the set of all individual triangles, and similarly the notion of a general quadrangle, pentagon, etc. They even went further and created the notion of a general polygon, the number of sides of the polygon being unspecified, and even variable.

This latter notion was already an abstraction from an abstraction, but it was a relatively easy one. It would have been much more notable and original if the Greeks, by suitable abstractions, had created the conception of a space of more than three dimensions, say of four dimensions, or, even more comprehensively, of any (unspecified) number of dimensions. But this they most decidedly never did, or ever remotely foreshadowed. Now, the statement just made about the Greeks might be objected to on the ground that the Greeks could not possibly have envisaged a multi-dimensional space for the following reason. The introduction of some types of spaces of several dimensions was effectively begun by Lagrange in his *Méchanique analytique* in the second half of the 18th century, when he introduced *generalized parameters* for mechanical systems with restraints. It was strongly dependent on the introduction of coordinate systems into ordinary space, and such coordinate systems gradually evolved out of the *Géometrie* of Descartes in the 17th century. The Greeks however, from whatever reasons, never introduced coordinate systems. Thus, one might object, it does not make sense to reproach the Greeks for not having introduced multi-dimensional geometry, if they happened not to be in possession of a device which is an indispensable pre-requisite for it. To this potential objection I would like to make two rejoinders.

First, from our retrospect, it is incomprehensible and "inexcusable" that the Greeks did not indeed introduce coordinate systems into their space. What is much worse, and much more inexcusable, they were hardly able to introduce a coordinate system on the (one-dimensional) time-axis, that is, in other words, to introduce a common calendar for general use. The citizens' chronology of the Greeks of the classical period is not a testimonial to their greatness.[44] For instance, Thucydides was a very eminent

[44] See the article "Aera" by T. W. Kubitschek in Pauly-Wissowa, *Real Encyclopadie,* Vol. I, Part 1 (1894), beginning with col. 606.

historian, and before him there had been a galaxy of
scientist philosophers—Thales, Pythagoras, Anaximander,
Anaximenes, Hecataeus, Heraclitus, Xenophanes, and
Parmenides, and others—whose names glitter from the
pages of history books. But between them they did not
generate enough spontaneity for the construction of a
usable calendar, so that there would be no need for
Thucydides to have recourse to his notorious "summer-
and-winter" calculus of dates, than which there is nothing
more embarrassing in the entire extant corpus of Greek
writing. When Thucydides wants to fix not only the season
but even a given month of a given year he has to verbalize a
monstrosity like this: "When Chrysis had been priestess at
Argos for 47 years, Aenesius being then ephor at Sparta,
Pythodorus having yet four months left of his archonship at
Athens" (Thucydides, ii, 2, I) = April, 431 B.C.[45]

Secondly, the Greek failure to conceptualize the notion
of space reaches back into the early stages of the notion, in
which the absence of coordinate systems would not yet be
of consequence. Pythagoreans, and certainly Plato in the
dialogue *Timaeus,* had a clear and mathematically satisfactory awareness that the space of events, that is, our
so-called Euclidean space, has three dimensions, that a
surface is a figure having two dimensions, that a line is
one-dimensional, and that a single point is, in our parlance,
zero-dimensional. Furthermore, the bulk of Euclid's geometry, namely Books 1–10 of the *Elements,* takes place in
a two-dimensional space, and Euclid is much more at ease
in this space than in the three-dimensional space of Books
11–13. Yet neither Euclid nor even Archimedes clearly
conceptualized this two-dimensional space of "ordinary"
geometry into a special entity in its own right and of its
own standing. It somehow always remained, conceptually,
a planar figure (*epipedon*) of relative existence only, that
is, one which was somehow immersed in the three-

[45] See "Chronology, Greek," in *Encyclopedia Britannica,* 14th
edn., V, 658d.

dimensional space of totality. The three-dimensional total space itself, the one and only space they were aware of, was not even properly conceptualized; as a mathematical datum it was not conceptualized to the stage at which it would receive an independent mathematical name of its own.

May I also observe that Greek natural philosophy in general was under the sway of a certain feature of anthropocentric sphericity, and that the failure of Greek mathematics to install coordinate systems can be linked to this feature. Lucretius, when arguing the unboundedness of the world in *De Rerum Natura* (Book I, lines 951–983), envisages the physical universe not, for instance, as a cube whose side is unbounded, but as a sphere whose radius is unbounded. In the center of the Lucretian sphere is a perceiving human mind, and Lucretius states expressly that, due to the infinitude of the radius, it is possible to shift the center of the sphere from any point in space to any other point. In the case of an unbounded cube the possibility of freely shifting its center suggests Cartesian coordinates, but not so in the case of an unbounded sphere. At best, the sphere would suggest so-called polar coordinates, but the calculus of such coordinates is indeed more than Archimedes and Apollonius could have mastered mathematically. However, the 17th century was not much better equipped for this analytical task, but it started out on the introduction of coordinates nonetheless, so that all this does not "explain" why Greek mathematics, in its strength, did not somehow somewhere crack the confines of this sphericity and open a pathway toward something new and fresh, for itself, and for Greek knowledge in general.

The efficacy, in mathematics, of abstractions from possibility and, conjointly with this, of abstractions from abstractions, reveals itself most manifestly in so-called algebra, if, in the present context, we understand by algebra the topic of calculatory operations with symbols. For instance, if we take ordinary rational numbers, that is, quotients of integers, as an actuality, then the general real numbers, which are limits of rational numbers, may be viewed as abstractions from *actuality* or, alternately, as one-step abstractions from reality. This given, the concept of a complex number $a + bi$, $i = \sqrt{-1}$, in which a, b, are real numbers, is an abstraction from *possibility,* inasmuch as it is *possible* to extend to such objects the basic arithmetical operations of addition, subtraction, multiplication, and division, in a suitable manner. Also, even if one assumes that complex numbers are only a one-step abstraction from real numbers, then, in order to reach them from rational numbers, it would take at least two successive abstractions. Next, if one proceeds to quaternions, one has to make at least one further abstraction, and the additional abstraction involved is certainly a very exacting one, inasmuch as the law of commutativity for multiplication no longer holds. General Cayley numbers require a further abstraction, and so on. And it is most important to keep in mind that what makes the advancement on this ladder of abstractions meaningful is the possibility of adapting and reinterpreting certain operational properties from any one abstraction to the next higher one. Abstractions without corresponding operations are mathematically pointless.

The Greeks rarely ventured beyond abstractions from actuality, in mathematics, in natural philosophy, and in other philosophy; and we wish to state, with emphasis, that the absence of algebraic symbolism from Greek mathematics was a symptom of a general Greek limitation, and not the cause of it. All Platonists notwithstanding, this gross deficiency in Greek "operational" thinking makes Plato's "idealist" epistemology entirely unsuited to deal with the torrent of abstractions from possibility, which is the hallmark of our time. It is illusory to presume that Platonic idealism can cope, interpretatively, with quantum field theory of today, even if Heisenberg himself should say so.

Georg Cantor's notion of a mathematical aggregate or

set, which ushered in a new era of rational thinking, is conceptually very direct; there is hardly an escalation of abstraction involved, and the Greeks were in fact constantly driven towards it. However, this notion comes to life only in an atmosphere in which creative operations with mathematical aggregates can prosper; the Greeks, not being inspired to produce the atmosphere, were not able to actually conceive the notion. For instance, in Plato's early dialogue *Charmides,* which deals with *sophrosyne* (= self-control, or so), Socrates elicits (166 C–E) from Critias the description that *sophrosyne* is an *episteme* (= knowledge) of other *epistemai,* including its own self; and Socrates insists on adding, as a further element, a species of negation-of-its-own-self (= ignorance). Now, if Plato had had the barest elements of set theory, *sophrosyne* might have become a set whose elements are: (i) itself, (ii) a bizarre anti-itself, and (iii) some other suchlike elements,[46] and this might have opened to Plato the wonders and paradoxes of set theory, which would have dwarfed the dialectical sterilities of the dialogue *Parmenides.* Also, Plato (169 A) even coined the momentous conception of knowledge of knowledge (*episteme epistemes*), which, set theoretically, would have become the Zermelo-like aggregate consisting of one element only, namely of "itself." Perhaps in imitation of this latter conception,[47] Aristotle introduced, in *Metaphysica* (Book XII, Chapter 9), as an attribute of God, the breathtaking notion of "thinking of thinking" (*noesis noeseos*), which, as a conceptualization,

---

[46] For information and references see T. G. Tuckey, *Plato's Charmides* (Cambridge University Press, 1951); and Klaus Oehler, *Die Lehre vom noetischen und dianoetischen Denken, bei Platon und Aristoteles,* Zetemata Monographs, No. 29 (Munich, 1962), pp. 123ff.

[47] For Aristotle's indebtedness to Plato in this case, see Paul Shorey, "Aristotle's *De Anima,*" *American Journal of Philology,* Vol. 22 (1901), pp. 149–164, especially pp. 160–164; also, Oehler (n. 46), pp. 201ff.

captivated commentators from Theophrastus to Aquinas.[48] This conception did enter into the fabric of our thinking, penetratingly so, but it did so non-mathematically, and none of the commentators themselves dared to break the Aristotelian seal of transmundanity in which the conception had been handed to them.

On the face of it, modern mathematics, that is, mathematics of the 16th century and after, began to undertake abstractions from possibility only in the 19th century; but effectively it did so from the outset. At first these abstractions, even when implied, were not very venturesome, and they were then always made for major purposes. Only the 19th century began to make them for the sake of making them; in the 20th century they have become very commonplace, as evidenced by the clamor for introducing them into school curricula under the motto of a "new" mathematics. Sometimes abstractions from possibility are performed to excess, as when there are, in a context, more abstractions than statements that link them, but the basic privilege of making them cannot be curtailed.

There seems to be a growing demand for abstractions from possibility in ever wider areas of organized knowledge, especially in natural science and social science, but also elsewhere; and it is this which makes it appear as if a certain mathematization of organized knowledge in general were in progress. In physics such abstractions, which were initiated in Newton's *Principia,* are beginning to abound almost as much as in mathematics. But mathematics continues to be the main repository for abstractions from possibility, old and new; and the "purer" the

---

[48] O. Hamelin, *La théorie de l'intellect d'après Aristote et ses commentateurs* (Paris, 1953); also, J. Tricot, *Aristote, la métaphysique,* Nouvelle édition (Paris, 1953), Vol. 2, Commentary to Chaps. 7 and 9 of Book XII, especially pp. 672–674 and 701–704. For axiological observations see W. J. Oates, *Aristotle and the Problem of Value* (Princeton University Press, 1963), pp. 114, 235–236.

mathematics, the purer the paradigms of such abstractions in it. It is this which makes many sciences seek affiliation with mathematics; they want abstractions of this kind much more than the mathematical techniques themselves; and, in the due course of developments, the "humanist" aspect of our civilization will only gain from these tendencies, rather than be obscured by them as some humanists seem to fear.

CHAPTER 2

HOW HISTORY OF SCIENCE DIFFERS
FROM OTHER HISTORY

THERE ARE differences of consequence, we contend, between history of science and other history, mainly general (that is, politico-social) history, but also history of literature, and even history of philosophy proper. The differences are differences of degree only, but they are significant. In detail, we will be mainly concerned with history of basic developments in mathematics and physics, but our observations will usually apply to larger areas of history, although the scope of the larger area may vary with the context. We assert as follows.

(1) History of science is younger, considerably younger than general history, and in historiographic analyses these differences of evolutionary age and maturity have to be taken into account.

(2) In history of science the relation of history to source material is an exacting one. The "Great Books" in science are at one and the same time events of science, source material for history of science, and frequently even themselves histories of science.

(3) In a certain qualified sense, especially with regard to basic or general developments, which are of interest to the public at large and not only to specialists in particular areas, the subject matter of history of science is, in volume, relatively small and has a relative scarcity of expandable detail. Therefore, in history of science, books of general narrative and over-all interpretation are, through no fault of theirs, exposed to the danger of becoming monotonous by repetition, or inflated by invention; whereas in general

history the need of broad-gauged narratives, and of intimate studies, is inexhaustible.

(4) The phenomenon of medievalism is much more pronounced and even more vexing in history of science than in virtually any other history. The all-important "rehabilitation" of the Middle Ages, which has been in progress since the first decades of the 19th century, is at least as important and illuminating in history of science as in any other history. But because of a relative scarcity of expandable material the peculiar danger of over-romanticizing the Middle Ages is, in history of science, even greater than in other history, and must be guarded against even more vigilantly.

(5) Next, unidirectedness of developments, irreversibility of advancement, and unlikelihood of recurrence of identical historical situations play a more incisive role in history of science than in general history. Science does change, irreversibly, not only in its *materia* but in the very nature of its *materia;* and more than in any other history, the past discloses itself in the future. This leads to the following problem with which history of science, especially history of leading ideas in science, has constantly to contend. On the one hand, due to the irreversible changing of science, it is doubly important to "fix" past events within their own past settings and against their own past backgrounds. But on the other hand, due to a continuing revealing of the meaning of events of the past in their unfolding in the future, it is frequently tempting, and almost irresistible, to appraise a scientific event of the past by what has become of it in a textbook of the future, and even in an up-to-date textbook of our today, that is, of any "today" in which the appraising is being performed. This does not mean to say that there are no durable findings in history of science, whose durability is impervious to the dating of the textbook by which to judge them. But we do mean to say that frequently the aspect of durability of a

finding in history of science is tinged by the same shadow of non-finality by which findings in science per se generally are. And this makes it the harder to distinguish the durable from the ephemeral.

(6) The most telling manifestation of unidirectedness and irreversibility of developments in science is the continuing and relentless systematization and mathematization of science. In consequence of this, the traditional feature of "spectacularity" and "wondrousness" of scientific phenomena will recede ever more. And it will become ever more difficult to uphold a substantive distinction between a history of science which likes phenomena of science to be unusual, and a philosophy of science which wants their attributes to be eternal.

(7) Yet, in spite of such pronounced differences between history of science and other history, searching critical appraisals of the art of historiography of science cannot be undertaken in isolation and separateness. Somehow all historiography is interconnected, and a significant development in historiography of science may have been activated by parallel developments in fields of history other than its own. A major illustration of this is the unfolding of the method of "Continuism." [1] In very brief, Continuism is a presumption that any event of today must be a reaction, concurringly or opposingly, to some directly preceding event, which must have taken place yesterday. In history of science Continuism began to come forcibly to the fore around 1900 in the voluminous work of Pierre Duhem, but it was in fact an old method from history of Greek philosophy which had been introduced by Aristotle.[2] In the course of the 19th century the method was greatly expanded and systematized by classicists, and Duhem put

[1] We took the term Continuism from Agassi, who speaks of "the theory of continuity." See J. Agassi, "Towards an Historiography of Science," *History and Theory*, Beiheft 2, 1963.
[2] See below, section 8, "Progress and Continuity."

it fully into operation for history of science at large, especially for the history of science in the Middle Ages. Continuism is somewhat prone to offer historical reconstructions without documentation; occasionally it is even doubtful whether there was an original "construct" at all to "re"-construct.

(8) Furthermore, it is important to realize that progress in the art of history of science is the outcome not only of developments in this art qua art, formally, but also of developments in science qua science, substantively. Thus, the method of "Conventionalism" [3] of Henri Poincaré, as an innovation in historiography of science, was the echo of "analytical" developments in philosophy of the foundation of mathematics; and, when it arrived, it did not arrive "too late," as sometimes bemoaned, but at a proper time, that is, at a proper stage of evolutionary developments. Popper [4] and Agassi [5] are right in asserting that before the advent of Continuism and Conventionalism the prevalent method in history of science was the "Inductivism" of Francis Bacon, and that Inductivism had, and still has, its shortcomings. Inductivism in history of science is, to begin with, the presumption that science advances gradually from observation to theoretization, and that it does so univalently, forcibly, and unerringly. This presumption is certainly "naïve," and it is usually accompanied by other presumptions which are equally naïve, as, for instance, by the presumption that there are ways in science of deciding between right and wrong, absolutely so, and that an experimenting or observing scientist can report on facts "faithfully" without, at all, rendering an "opinion" on them. Nevertheless, when appraising the work of a historian of science, the bare fact of his having been, or be-

[3] See below, section 10, "The Mathematization of Science."
[4] K. R. Popper, *Conjectures and Refutations* (New York, 1962), pp. 137ff., and other passages (see index under "induction," etc.).
[5] See n. 1.

ing, an inductivist ought not to be held critically against him as such. I venture to speculate that if Isaac Newton had composed a history of science on a systematic scale, it would have been a fully inductivist one.[6] It might have contained some strands of continuism, as are already detectable amid his historical remarks in the *Principia*. But it could not yet have been conventionalist to a recognizable extent. Conventionalism presupposes a certain degree of articulation of philosophical insights into, and a certain attitude toward, the nature of mathematics which only the 19th century brought forth organically. The 17th century was by no means ready for it, and, great and forward-directed as Newton was, in questions of the articulation of basic structure and philosophy of mathematics proper, he was in no wise ahead of his times. Thus, for instance, Newton was unquestionably and for the ages the true founder of the application of the infinitesimal calculus to mechanics and physics, and yet, much as he may have tried, he was not able, to the end of his days, to define with some measure of rigor the conception of a derivative of a function, let alone of a second derivative of it. And Newton was certainly no exception. Neither Euler nor Lagrange, long after Newton, could really define a derivative, although Euler introduced coordinate mathematics into mechanics, and Lagrange, about 100 years after Newton, was in a sense the actual creator of the Theory of Functions. Another 50 years had to pass, which is quite a long time, before something resembling our textbook conceptions of the calculus began to take shape.

1. SLOWNESS OF THE RISE OF HISTORY OF SCIENCE

History of science, as many another non-general history, has become an organized discipline, with a status of its own, much more slowly than history in general.

[6] Compare F. E. Manuel, *Isaac Newton, Historian* (Harvard University Press, 1963).

For history of science there is nothing extant from antiquity that would compare, in intellectual strength and organization, with the works of Herodotus, Thucydides, Xenophon, Polybius, Tacitus, and others, nor even with the historical works of the Old Testament, however much the latter may be affected by theocratic pragmatism. There are two important sets of "Lives" from Greek antiquity, the *Lives of Public Affairs Men* by Plutarch (A.D. 46–120), which belongs to general history, and the *Lives of Eminent Philosophers* by Diogenes Laertius (A.D. 200–250), which belongs to history of science and natural philosophy. And it so happens that qua history the work of Diogenes is greatly inferior to that of Plutarch.

Pliny's *Natural History* was not intended to be, and did not become, a history of developments, and apparently there were no evolutionary Natural Histories in antiquity or the Middle Ages. But there were Greek histories of individual scientific disciplines, and a well-attested instance is that of a history of geometry by Eudemus, a pupil of Aristotle. Proclus (A.D. 411–485), in a commentary on Euclid's *Elements,* has a chronology of the rise of Greek geometry,[7] which is apparently taken from, or based on, the work of Eudemus. This lengthy passage has been analyzed to exhaustion, not because it is inspiring history, but because it is all one has. Another lengthy passage, this one purporting to come nearly verbatim from Eudemus, was preserved by Simplicius (6th century A.D.). It deals with the quadrature of the circle, and mainly with the work of Hippocrates of Chios from the 5th century B.C.; however, it somehow is more in the nature of an early text in geometry itself than in history of geometry.[8] This history

[7] (i) T. L. Heath, *The Thirteen Books of Euclid's Elements,* 3 vols. (Oxford University Press, 1908), 1–6; (ii) T. L. Heath, *History of Greek Mathematics,* 2 vols. (Oxford University Press, 1923), I, 118–121.

[8] See in the Loeb Collection, Ivor Thomas, *Greek Mathematics,* I, 235–253.

of Eudemus was apparently an "indispensable" source book in later antiquity, but it is difficult to make out whether Eudemus crossed the line between chronicling-and-reproducing and doing "genuine" history.[9]

There exist several essays by Aristotle on history of science, in the form of historical introductions to full-length treatises on special areas of science or scientific knowledge. Such an introduction frequently occupies Book I of the treatise; see for instance his *Metaphysica, Physica, De Caelo,* and *De Anima.*[10] These historical surveys in science by Aristotle may have been the first of their kind to deserve the name, but they are all slanted and even biased, and there is a great dearth of other documentation against which to check them indisputably. One must not assert, however, that this slanting was an incurable failing of Aristotle. He could do history beautifully when he cast himself into the role of historian, as he did in his *Athenian Constitution,* and as he also did in his *Politics,* when giving accounts of various social orders and political constitutions, past and present. But in his scientific treatises Aristotle always casts himself in the role of scientist and not of historian, and the historical notices about others of the past are presentations of views of predecessors of his qua predecessors and forerunners, and not of views of previous scientists in the role of actors in the continuing play on the stage of science. Ever since Aristotle, working scientists have been behaving in this way, that is, showing biases and slanting of the Aristotelian brand when referring to the work of others, be they predecessors or contemporaries. They never "quote" others with a faithfulness and attention to details that would satisfy philosophers, and they are exasperating in the manner in which they credit

[9] Eudemus may have also written a history of astronomy. See, for instance, G. S. Kirk and J. E. Raven, *The Presocratic Philosophers* (Cambridge University Press, 1957), p. 80.

[10] It is undecidable whether, before Aristotle, works under the general title "On nature" included histories.

achievements to predecessors with ever varying degrees of accuracy, or rather inaccuracy. Yet on the advancement of science itself, as against *history* of philosophy and science, Aristotle's "bad example" of extracting a "gist" from developments of the past, and then proceeding from there, had a very good effect on the whole.

One may even question whether it is rewarding to be conscientious in reporting on predecessors. The *Almagest* (A.D. 150) of Ptolemy embodies historical retrospections of great importance, and much of what is known about the astronomer Hipparchus (born 190 B.C.) comes from the *Almagest*. Ptolemy's historical statements bear the mark of information about predecessors qua predecessors, but Ptolemy obviously strove for historical objectivity. He was extremely laudatory about Hipparchus—as Plato had been about Parmenides, say—and he was precise and detailed when describing and delimiting the achievements of Hipparchus. The little that has survived of the writings of Hipparchus in original happens to be of such an ordinary kind that no reputation could be built on it. But, as a "reward" to Ptolemy for his efforts in behalf of Hipparchus, "everybody" [11] says that Hipparchus (and not Ptolemy) was the greatest astronomer in antiquity, and some commentators give the impression of barely being able to control their suspicions that Ptolemy must undoubtedly have been inequitable to Hipparchus in something, somehow.[12]

In the Middle Ages, statements pertaining to history of

[11] But not O. Neugebauer, *The Exact Sciences in Antiquity,* 2nd edn. (Brown University Press, 1957).

[12] Even the eminent T. L. Heath cannot quite control himself. In his *Greek Astronomy* (London, 1932) he does, on page lii, only say that Hipparchus was "perhaps" the greatest astronomer of antiquity. But, on page liii, when speaking of Ptolemy he says: "It is questionable whether Ptolemy himself added anything of great value except a definite theory of the motion of the five planets, for which Hipparchus had only collected material in the shape of observations made by his predecessors and himself."

science occur, perhaps, in commentaries on Aristotle, interspersed among interpretations of Aristotelian pasages which are commented upon.[13] In the 16th and 17th centuries, notices on predecessors qua predecessors are spread over the works of Copernicus, Kepler, Bacon, Galileo, Descartes, Newton, and others. Prominent among these are utterances against Aristotle and Aristotelians. They are not historical judgements which happen to be unfavorable but, for the most part, ritualistically performed bitter condemnations of predecessors for having erred and misled. After that, regular histories of special sciences, mainly exact sciences, began to be composed in the 18th century. Leading among these [14] were Montucla on mathematics,[15] Priestley on electricity,[16] Lagrange on mechanics,[17] Bailly on astronomy,[18] and, above all, La-

[13] Even Thomas Aquinas in his *Commentary on Aristotle's Metaphysics* (translated by T. P. Rowan, Library of Living Catholic Thought, 2 vols., Chicago, 1961), in his observations on the historical Book I, does no more than give a very lucid exposition of the text as an aid for the serious but "ordinary" learner; but there is not much to arrest the attention of the research scholar.

[14] As pointed out to me orally by C. C. Gillispie, although formal histories of exact sciences began to be composed only in the second half of the 18th century, a measure of acknowledgement is due to Bernard le Bovier de Fontenelle (1675–1757) for his contributions to the analyses of proceedings in the *Histoire du renouvellement de l'Académie des Sciences* (3 vols., Paris, 1708–1722), and for his famous *Éloges* on academy members. (For an English language epitome see the edition by John Chamberlayne, *The Lives of the French, Italian, and German Philosophers,* 1717.)

[15] J. F. Montucla, *Histoire des mathématiques,* 2 vols. (1758; second edition in 4 vols., 1799–1802, reprinted 1960).

[16] Joseph Priestley, *History and Present State of Electricity* (1767); he also composed a *History of Discoveries Relating to Vision, Light and Colours* (1772).

[17] J. L. Lagrange, *Mécanique analytique* (1788) has important lengthy sections on history of the subject.

[18] J. S. Bailly (1736–1793); *Histoire de l'astronomie ancienne* (1775); *Histoire de l'astronomie moderne,* 3 vols. (1779–1782); *Traité de l'astronomie indienne et orientale* (1789). Also, *Lettres sur l'origine des sciences* (1777), and *Lettres sur l'Atlandide de Platon* (1779).

place, also on astronomy.[19] Afterward, in the first decades of the 19th century, standard histories of astronomy were composed by Delambre.[20] The historical part of the treatise of Laplace [21] was a non-technical essay for the general public, and an epitome of the work of Bailly was composed in a similar style.[22]

Astronomy is, next to mathematics, the oldest known organized (exact) science, and its hold on the imagining and thinking of *Homo sapiens* reaches back perhaps into the stages of myth-making. And the slowness of the rise of history of science is made manifest by the fact that, even in astronomy, "regular" histories were so tardy in coming, those of Laplace and Bailly being the first. Both these works were inductivist, sometimes adversely so. But this fact must not be allowed to detract from their outward merit of being "firsts," and from their inward merit of being first-rate work by first-rate men.

It is true that in most inductivist histories of science, even through the length of the 19th century, the settings were reminiscent, methodologically, of settings which had been prevalent in general histories up to the late decades of the 18th century, before Romantics introduced the conscious and systematic emphasis on rationalizing, or even glorifying, the (seemingly) irrational (in the past); and before post-Romantic scholarship, soon proliferating, began to spread and practice the conscious and systematic emphasis on scrutinizing and evaluating historical docu-

---

[19] P. S. Laplace, *Exposition du système du monde* (1796). *Oeuvres complètes,* Vol. VI, (1884); especially, Livre V, *Précis de l'histoire de l'astronomie.*

[20] J. B. J. Delambre (1749–1822), *Histoire de l'astronomie ancienne,* 2 vols. (1817); *Histoire de l'astronomie au moyen âge* (1819); *Histoire de l'astronomie moderne,* 2 vols. (1821); *Histoire de l'astronomie du XVIII siècle,* completed by C. L. Mathieu (1827).

[21] See n. 19.

[22] J. S. Bailly, *Histoire de l'astronomie ancienne et moderne* 2 vols. (Paris, V.C., 1805).

mentation not only with regard to a simple true-or-false, but with regard to various and varying levels of authenticity and significance. In this sense, the non-technical history of Laplace, and even more so the histories of Bailly, are indeed somewhat behind the times. But we wish to point out that earlier in the 18th century Montesquieu and Voltaire had composed works in general history which enjoy considerable renown, even though methodologically they do not yet bear the marks of sophistication which the Romantics began to impose. Due to the relative lateness in the rise of history of science, it is justified to judge the works of Laplace and Bailly by criteria which are not harsher than those usually applied to the works of Montesquieu and Voltaire. But if so judged, the works of Laplace and Bailly must be adjudged to have been very good indeed, whatever shortcomings their methodology may, allegedly, have had.

The first significant "universal" history of science, still very readable, came only several decades after 1800, and it was due to the author of divers works, William Whewell (1796–1866).[23] By general circumstances, the fact that a work like Whewell's was done as and when it was done is not remarkable in itself. Whewell was a classicist, and the history of Greek natural philosophy was to him a part of history of science, as a matter of course. Now, in a certain sense, history of Greek philosophy had been a topic of systematic scholarship since antiquity,[24] much longer at any rate than history of science as such. Hellenistic commentators made many contributions to this history, after their fashion. From the 16th century onwards, among classicists and philosophers, interest in the subject matter had been steadily increasing, and among contemporaries of

---

[23] *History of the Inductive Sciences* (1837); also, *Philosophy of the Inductive Sciences* (1840).

[24] J. E. Sandys, *A History of Classical Scholarship*, 2 vols. (1606–1908).

Whewell, some even older than himself, the level of these studies was quite advanced. One need only mention August Boeckh (1785–1867), who was both a historian *à la* Niebuhr and a pioneering classicist with good understanding for the need of "relativistic" appraisals of the evolutionary aspects of Greek natural philosophy.[25] In view of this it is not surprising to find that Whewell was able to perceive features of gradualness even in the "revolutionary" 17th-century age of science,[26] and in hindsight one might even wonder why he did not anticipate more than he did other methodological variants of historical "relativism," especially the "continuism" of Duhem and even the "conventionalism" of Poincaré, seeing that he was also familiar with rational mechanics.

In mathematics, the rise of its historiography was also inexplicably slow. The *Principia* of Newton (1686) gave reality to the Pythagorean vision of articulating science through mathematics, and the mathematics that went into the composition of the *Principia* was at least 2,300 years old, probably very much older.[27] But, when the *Principia* appeared, there was not a book around that told the story of this mathematics, not until, 70 years later, Montucla wrote his. The work of Montucla, fascinating as it is, does not attain to the grandeur of Gibbon's *Decline and Fall of the Roman Empire,* with which it was coeval. However, Montucla did not have his Suetonius, Eusebius, Procopius, Anna Comnena, etc. to fall back upon; there was not even a real chronicle deserving the name. From antiquity there was only a listing of some mathematicians and their doings

[25] For references see, for instance, J. W. Thompson and B. J. Holm, *History of Historical Writing* (1942), II, 156n. Especially, Boeckh's *Encyclopädie und Methodologie der Philologischen Wissenschaften,* edited by E. Bratuschek in 1877, which has a long section on Greek (natural) philosophy.

[26] Agassi (n. 1), p. 30.

[27] O. Neugebauer, "The Survival of Babylonian Methods in the Exact Sciences of Antiquity and Middle Ages," *Proceedings of the American Philosophical Society,* vol. 107 (1963), pp. 528–535.

in the commentary on Euclid by Proclus (A.D. 410–485). To judge by the listing of extant source material in the standard treatise on Byzantine literature,[28] the large corpus of Byzantine writings had virtually nothing to offer to Montucla. In the Middle Ages of the West, Roger Bacon (1214–1294) sang the praises of mathematics,[29] but, in truth, he hardly knew what mathematics is. Nicole Oresme (1330–1382) did know what mathematics is, and he was a fully active promoter of mathematics; but he pursued many divers intellectual activities, and, furthermore, his preoccupation with mathematics was not of the kind to make him want to survey its past. During the 16th and 17th centuries everybody was so busy "advancing the frontiers" of mathematics that nobody cared to know how it came into being. And even after Montucla laid the groundwork in 1758, it still took 120 years before a standard history of mathematics was inaugurated when the large treatise of Moritz Cantor (1829–1920) began to appear in 1880. But it should be noted that Montucla's manner of viewing mathematics as a part of civilization remains engrossing, in spite of the fact that it was a widespread mannerism of the 18th-century Enlightenment to view anything historical as a part of history of civilization at large.

In the 20th century, a "universal" history of science by W. C. Dampier-Whetham (first edition 1930) engendered much popular interest in the subject matter, and the large work of George Sarton (1884–1956), mainly bibliographical, is simply indispensable. But in the judgement of some younger historians of science, Dampier and Sarton are both badly tainted with inductivism, which in their case seems to mean that they believed that science is propelled

[28] K. Krumbacher, *Geschichte der Byzantinischen Literatur*, 2nd edn. (1897). On p. 626, Krumbacher seems irked because Moritz Cantor, *Geschichte de Mathematik* (3rd edn., I, 488–517), calls Byzantine mathematics "degenerate" ("Die Griechische Mathematik in ihrer Entartung").

[29] Mainly in his *Opus Maius*.

by a breed of heroes who never err or fail. Perhaps this was so; some such attitude was indeed theirs, and it was indeed a limitation. But in the case of Dampier I venture to suggest, in a lighter vein, that by the disparity between the ages of history of science and general history, the book of Dampier is actually coeval with the *Chronicles of Knightly Heroes* by Jean Froissart (1337–1400), and that such a comparison would not be injurious to either. And about Sarton I wish to say that after his many-volumed *Introduction to the History of Science* began to appear in 1927, I used to peruse cursorily the entries on mathematics and was never struck by a statement that was false or vapid.

Finally I wish to observe that within present-day New Directions in history of science some of the avant garde positions are held with more verve and militancy than called for, and the prime case is that of the phlogiston theory in the chemistry of the 18th century, which held sway until oxygen and its properties became familiar. The theory was nothing to be very "ashamed" of; in fact from the hindsight of present-day particle physics one might say that, in its last stages, phlogiston was a would-be "anti-matter" to oxygen, and it was so very much "anti" that even its gravitational mass was negative. However, from whatever reasons the straightlaced Victorian chemists of the 19th century viewed the bygone phlogiston theory as an unprofitable aberration, and the memory of it as an embarrassment. But, as if in a countermove, recent evolutionary approaches have begun a process of rehabilitating the phlogiston interregnum by elaborating on specific advances which were achieved in chemistry during that period, directly or contingently. This in itself is legitimate and illuminating, if not carried too far. However, regrettably, the rehabilitation is sometimes promoted with unnecessary disrespect for the views of those who dissent. It is true that the old-fashioned outright condemnation of the phlogiston theory was lacking in a certain kind of

historical imaginativeness and pliability. But I make bold
to assert that the underlying rationale of condemnation
retains a measure of justification which no effort at
rehabilitation will erase. And it is refreshing to find the
old-fashioned view restated, unconcernedly, and with very
good historical sense, in a fairly recent book by John
Read.[30]

2. TRENDS AND INDIVIDUALITY

Before proceeding with our subject matter proper in the
next section, we will in the present section insert several
remarks, some of them referring to inductivism. The
remarks will not be closely connected with each other by
an overt nexus. Cumulatively, however, they will suggest
that although there are indeed differences between history
of science and other history, yet the nature of the
differences is sometimes elusive and the determination of it
a challenge.

Nowadays it would be considered grossly naïve to
presume, flatly, that progress in science is foreordained,
irresistible, and unequivocally recognizable; that, in science, right is right and wrong is wrong, and that what is
right is true and what is wrong is false, absolutely so; and
that in science there simply must be a way of choosing
between true and false. Nevertheless, presumptions like
these used to have wide currency, even through the whole
length of the 19th century.[31] However, they were not at all
confined to science. Analogous presumptions, even ramp-
ant ones, occurred in general history too. R. G. Collingwood [32] reports, with contempt, on a Late Victorian book

[30] *Through Alchemy to Chemistry* (London, 1957).
[31] May I recall that the author of Deuteronomy 13 also takes
it for granted that there is a means, an infallible one, of dis-
criminating between a true prophet and a false prophet.
[32] *The Idea of History* (Oxford University Press, 1957), pp.
145–146.

in 19th century history [33] which, in the words of Collingwood, depicts "that century as a time of progress from a state of barbarism, ignorance, and bestiality which can hardly be exaggerated, to a reign of science, enlightenment, and democracy." Collingwood quotes from the book an excerpt, than which there could be nothing more grotesquely progressivist in a book on history of science, no matter how inductivist.[34]

On the other hand, a statement in history of science, a legitimate and plausible one, may become entirely distorted when lifted out of its context and ingenuously applied to history in general. Take for instance the following statement: [35]

> The inductive style of presentation, according with which one records facts faithfully, and omits opinions —or perhaps mentions them briefly in the final paragraph of one's report—was suggested by Bacon and Boyle.

It seems to me that, in general history, very long before Bacon and Boyle, this style of presentation was "suggested" by Thucydides and made him immortal.

Sometimes, however, it is appropriate to transfer a judgement from history of science to general history, even if first impressions warn against doing so lest it be done hastily. The inductivist belief that in science a theory or statement of the past was either right or wrong, or either true or false, may also be expressed, reproachfully, in this way: that inductivism indulged in sharp contrasts between black and white with no intermediate colors allowable. Now this reproach may also be applied, quite justifiably, to Gibbon's *Decline and Fall,* especially since Gibbon had a

[33] Robert Mackenzie, *The 19th Century—A History* (1880).
[34] See n. 32.
[35] Agassi (n. 1), p. 93.

pronounced tendency to blacken as a historian what he did not like as a person. Also, Gibbon's attitude towards history falls into the general tableau of Enlightenment, and one may say that, in the 18th century, Enlightenment in the "humanities" closely corresponded to Inductivism in "science." But we wish to point out that precisely because Gibbon's limitations seem to stem from a general trend which was prevalent during his epoch it is customary not to condemn him roundly for this kind of limitation, but to make allowances for it, and to acknowledge his achievements, which were durable, rather than scorn him for his failings, which were time-bound.

As a matter of fact, a variant of the "reproach" against Gibbon may even be leveled against Theodor Mommsen, even though professional historians but rarely dare, for the record, to suspect Mommsen of having been other than impeccably Ranke-ish. Mommsen's *Roman History* was intended to be a work in five volumes, but Volume 4 was never written. Volume 3 ends with Pompey and Caesar, and Volume 5, the famous one, deals with Roman provinces under the emperors; but Volume 4, about the Emperors themselves, is missing. Mommsen was an untiringly prodigious worker who would not have failed to complete an undertaking like this from lassitude, say, and there is in print no satisfactory explanation why Volume 4 was never written. But I have it in my memory that an "oral" explanation has been as follows. In Volume 3, Mommsen painted Pompey blacker than black and Caesar whiter than white; and (being some kind of "inductivist") he had no suitable color with which to paint Augustus, with whom the missing volume would have had to begin. To paint him in a "colorless" fashion would, of course, have been incongruous.

An entirely different remark refers to the question to what extent it is meaningful to derive a working scientist's *individual* creativity in his professional work from a

[75]

general cultural background, or from a "deliberate" philosophical posture of his, or from internal debates of his
about such postures. It can be very misleading to do so in
a glib and pat manner, by preconceived formulas; especially since scientists (even scientists!) are as fully endowed with individualities, caprices, idiosyncrasies, and
temperaments which defy subjugation to a *Zeitgeist* as
anybody else.

It is a fact, for instance, that Euclid, Archimedes, and
Apollonius, in their *formal* mathematical works, observed
a strict rule of removing "all traces of scaffolding" [36] by
which a result had been arrived at, or a demonstration had
originally been built. Also there are similar stylistic
rigidities in Greek tragedy and even in Thucydides. It is
possible, and even likely, that appropriate stylometric
investigations might produce parallelisms of structure
between these various formalisms, but it would nevertheless be most misleading to summarize all such formalisms
into one over-all formula about Greek civilization, say.

A Marxist might find it fitting to say that an Archimedes
and Apollonius showed this stylized fashion because they
were disinterested in spreading their knowledge among
broad strata of citizenry, but this would hardly explain why
an Aeschylus, who undoubtedly wanted to reach a large
audience, wrote in a language which, for all its magnifi-
cence, was so stylized that the text has become infested
with corruptions simply because Hellenistic actors and play
directors were unable to understand the intended meaning
of words and sentences (Homer has almost no corruptions!). Next, scholars in the 16th and 17th centuries
publicized their achievements in charades, anagrams,

[36] As Clerk Maxwell once put it, in another context. We note
however that a considerable amount of "scaffolding" is showing
in "The Method" of Archimedes which was discovered by L. Heiberg in 1906. See Ivor Thomas, *Greek Mathematics* (Loeb Collection), II, 221ff.

acrostics, etc., but I do not know what kind of philoso-phemes about trend one could make in this case. Kepler, in his *New Astronomy,* hid his first two planetary laws inside a most prolix narration of failures and successes, and it took the genius of Newton to bring them to light. But Newton himself always gives the impression of telling less than he might, even though the *Principia* is a long book, and the *Queries* to the *Opticks* are intentionally revealing. The mathematical work of L. Euler (1707–1783) is open-minded and the largest on record, and yet it never sounds prolix. J. L. Lagrange (1736–1813) speaks much less freely, although he does not actually hold anything back. C. F. Gauss (1777–1855), however, whose work is quite large, always conceals something, or so it seems. A. Cauchy (1789–1857), on the other hand, gives the impression of pouring out everything in an unending stream,[37] but in some of his principal work he can be controlled and sometimes even summary.

Finally there was a great contrast in temperament and working method between Ampère (1775–1836) and Faraday (1791–1867), which attracted much attention and elicited much comment. Thus Maxwell said [38] that the method of Ampère "though cast in inductive form" could not have been the one by which he originally discovered his law of action, whereas Faraday shows not only his successful experiments but also his unsuccessful attempts, thus giving encouragement to a reader "inferior to him in inductive power." Agassi [39] seizes on Maxwell's use of the

[37] The Académie des Sciences de Paris, when deciding on 13th of July 1835 to initiate its "Comptes Rendus," immediately adopted limitations on the amount of printed space to be available to its various categories of members (see *Proces-Verbaux des Séances de l'Institut jusqu'au mois d'août 1835,* Vol. 10 (1922), pp. 728–731 and p. 756). Now, E. T. Bell, *Men of Mathematics* (New York, 1937), p. 287, suggests that this was done against Cauchy.

[38] Agassi (n. 1), p. 90–94.

[39] See n. 38.

word "inductive" to conclude that Faraday was consciously anti-inductivist in his philosophical posture; and Agassi seems to assert that Faraday was consciously waging a struggle against inductivism in his thinking and that his scientific breakthroughs and decisive achievements were the outcome of this struggle and the measure of his capacity of winning this struggle. Now, it is hard to refute such an assertion directly; but there is something forced and ill-motivated about it. From what is known about Faraday and Ampère as persons, it appears that by everyday standards of temperament Faraday was an amiable and agreeable person, whereas Ampère was a much more formal person, with whom it was much less easy to associate. It is therefore possible and, in fact, very suggestive to explain the differences in their working methods from differences in temperamental disposition, without making very hypothetical assumptions about differences in philosophical posture. Also, as pointed out to me by T. S. Kuhn, at the time of Ampère and Faraday the organization of the pursuits of science was much less formal in Great Britain than in France, and the difference of temperament between Ampère and Faraday may have been accentuated by differences of their regional backgrounds.

It is true that in the 20th century, scientific reports, especially on experimental investigations which emanate from institutes and laboratories tend to have the form of a uniform manner of communication in which the individuality of temperamental disposition is submerged in some kind of uniformity of posture. However, this is due not to some trend in philosophy, but to the growing uniformity of organization, in which "chief investigators" have to depend on the collaboration of research groups, of teams of assistants, and of crews of technicians. In keeping with this, the communications tend to have a standardized form of reporting, and frequently one has the impression that the reports are extracts from notebooks and logbooks of a standardized kind. But all this only reflects standards of our technology, not postures of our philosophy.

Finally I wish to comment on an interesting situation involving Isaac Newton. It is a fact that Newton did not concern himself with the concept of energy as we have it in physics today, and that, specifically, no version of the law of conservation of energy is traceable to the *Principia;* Newton failed to do this even though, to his durable credit, he finally and firmly set down the concept of the translational momentum, together with its law of conservation (if there are no forces), and then, *implicitly,* even introduced, by abstraction from the second planetary law, a very first version of the law of conservation of the angular momentum (in general central forces). It is a further fact that Newton's contemporaries Huygens and Leibniz were aware of (and were publishing about) the notion of energy and some of its properties. To some observers this further fact makes Newton's silence in the matter of energy doubly puzzling and even quite disturbing. There seems to be a suspicion abroad that this silence stemmed from a deep-seated antagonism of Newton against Leibniz, which was in part philosophical and in part even personal, and that it was this peculiarly motivated antagonism which dictated to Newton's scientific interests and creativity that the subject matter of energy was to be avoided.[40]

The puzzle was compounded by a publication of the Marxist interpreter B. Hessen.[41] In it he seems to assert, in detail, that in Newton's make-up and background there was some kind of socio-economic motivation which conditioned him, quite unequivocally, and almost mischievously,

[40] For references see: E. N. Hiebert, *Historical Roots of the Principle of Conservation of Energy* (Madison, 1962); also Adolf Kneser, *Das Prinzip der kleinsten Wirkung von Leibniz bis zur Gegenwart* (Berlin, 1928).

[41] B. Hessen, "The Social and Economic Roots of Newton's *Principia,*" a reprint from, and also contained in, the cooperative volume: *Science at the Cross-Roads* (London, 1931).

against finding, in physics, the concept of energy and the law of conservation of energy; and Hessen seems to maintain that Newton's "ban" on these discoveries was so strong and effective that it took over a century and a half until, in the middle of the 19th century, the discoveries could finally be made. Some specialists in English history of the 17th century took Hessen's accusations seriously, and G. N. Clark replied to them in full.[42] Also, according to Clark, it is Hessen's interpretation that Newton failed (or perhaps "refused") "to anticipate the views of Friedrich Engels on the interconnection and hierarchy of all forms of matter." [43]

Now, I wish to remark to the above that to myself there is nothing puzzling or disquieting in Newton's failure to discourse on energy, the simple explanation being that Newton did not need it and also was not ready for it, mathematico-physically, that is.[44] It is a fact that in the evolution of science since 1600 the conception of energy, in depth, evolved not from mechanics, but only later from physics proper; and the physics proper from which it evolved began to emerge only during the period between 1780 and 1820, whatever might have gone on before. Even in optics, the present-day structure of this discipline is that of Fresnel and not that of Descartes–Snell, Huygens, and Newton. As a matter of fact, the conception of energy, in its actual significance, came from the "non-mechanical" thermodynamics of Sadi Carnot and successors, and not

[42] G. N. Clark, *Science and Social Welfare in the Age of Newton* (1937), p. 62.

[43] It so happens that the term "matter" in Engels also subsumes "energy"; see Friedrich Engels, *Dialectics of Nature*, edited by C. Duff (New York, 1940), editor's footnote on page 27. Therefore, according to Hessen (via Clark), Engels had formulated the law of conservation of energy much more sophisticatedly than Mayer and Helmholtz, namely, immediately, as a 20th-century law of conservation of matter-cum-energy; which of course is preposterous.

[44] This point will recur in later contexts.

earlier from the work of Poisson on electro- and magneto-
static potentials,[45] let alone, from preceding work of Euler
and Lagrange, even if, from our retrospect, it was already
there, with versions of the law of constancy of energy too.
Furthermore, Newton was temperamentally not given to
publicizing leading ideas of his before brooding over them
for a considerable length of time and satisfying himself of
their eventual significance; whereas Leibniz, say, would
pre-announce intended forthcoming pursuits of his in a
relatively "quick" note, or even in a letter *à l'occasion*. In
sum, I think that Newton did not intervene in the subject
matter of energy because he had no contribution to make
that would have been worthy of his intervention.

Against this, one might argue that the mathematical
schema of "energy" is a simple scalar, whereas for
translational momentum, which grew on Newton's home
ground, it is the less simple notion of a vector, and in the
case of angular momentum, which he also anticipated, it is
even a skew-tensor. That is, if Newton could find conceptions whose mathematical representatives are vectors or
even skew-tensors, then he ought to have found the
conception of energy, whose mathematical representative
is a common scalar only.[46] To this I can retort that ordinary

[45] E. T. Whittaker, *A History of the Theories of Aether and
Electricity*, I (1951), 60–65.

[46] A *scalar* is a mathematical object, frequently of geometrical
origin, which is determined by a single number.

A *vector* is a mathematical object which is attached to a space,
and which requires for its determination as many numbers as
there are dimensions in the space. Thus, in the case of three-
dimensional space, a vector is determined by a set of three numbers

$$a_1, a_2, a_3$$

which are its *components*.

A *tensor* in three-dimensional space is determined by a 3-by-3
array of numbers, thus

$$a_{11}, a_{12}, a_{13}$$
$$a_{21}, a_{22}, a_{23}$$
$$a_{31}, a_{32}, a_{33}$$

The components

$$a_{11}, a_{22}, a_{33}$$

real numbers are also only scalars, and yet, whether by a coincidence or not, the "modern" mathematical foundation of real numbers was laid *only* in the middle of the 19th century, at about the same time that the conception of energy in physics was taking its deep roots. It is, of course,

---

of the array constitute its diagonal, the components

$$a_{12}, a_{13}, a_{23}$$

are said to lie above the diagonal, and

$$a_{21}, a_{31}, a_{32}$$

are said to lie below the diagonal.

A tensor is *symmetric* if an interchange of rows and columns does not change the values of the components. The requirement of symmetry does not restrict the components on the diagonal; but the elements above and below the diagonal are pairwise equal, thus,

$$a_{21} = a_{12}, a_{31} = a_{13}, a_{32} = a_{23}$$

Therefore, *in effect*, a symmetric tensor is determined by six numbers only, and not by nine numbers as a tensor in general.

A tensor is *skew-(symmetric)*, if an interchange of rows and columns replaces any component by its negative value. For the diagonal components this means

$$a_{11} = -a_{11}, a_{22} = -a_{22}, a_{33} = -a_{33}$$

Therefore these components are all zero,

$$a_{11} = a_{22} = a_{33} = 0$$

and the pairing of the components above and below the diagonal now has the form

$$a_{21} = -a_{12}, a_{31} = -a_{13}, a_{32} = -a_{23}$$

Therefore, *in effect,* a skew-tensor in three-dimensional space is determined by three components only, and it therefore also is in the nature of a vector.

In the 19th century, a vector which arises from a skew-tensor was called *rotational,* whereas an "ordinary" vector was called *translational.* In the electrodynamics of Maxwell, the electric force E is a translational vector, and the magnetic vector H is a rotational vector; which is in keeping with the facts that electricity is "usually" encountered in "currents," whereas magnetism "usually" occurs in "dipoles."

That a rotational vector ought to be viewed as a skew-tensor was, in a sense, already known to H. E. Grassmann in the middle of the 19th century; it was made part of the substance of physics in 1908 in the work of H. Minkowski on Special Relativity; and it was afterwards emphatically stated by Hermann Weyl in his book *Space, Time, Matter.*

The role of vectors and tensors in physics will be discussed, in general terms, in chapter 6, section 9.

extremely lop-sided to state, as Hessen did, that Newton failed to anticipate some species of view of Friedrich Engels', even if, literally, the statement cannot be refuted. However, it is proper to state that Newton failed to anticipate scientific conceptions which only the 19th century brought forth; and the conception of energy was one among these.

3. RELATIONS TO SOURCE WORKS

In political and social history, whatever the rigors of documentation may be, at least as far as the reading public is concerned, the purpose and effect of the writing of history is in the writing about human doings, events, and changes, and not in the writing about the documents themselves from which the events are known. In contrast to this, in history of literature (including bibles), the writing is in the main, though not entirely, about other writings, so that the historical books are usually about other books, namely about the poets' and writers' original books. History of science falls somewhere in between. In it, to a considerable extent, source books are themselves events and not only books by which to reconstruct events. Therefore, the textual integrity of the source texts has, to varying degrees, some aspects of the textual sanctity of the poets' works. But the situation is rather involved, as we will try to indicate by brief interlocking arguments.

Most of what is known about Greek "events" in the 5th century B.C. goes back ultimately to the works of Herodotus and Thucydides, which were themselves written in that century. These works, by themselves, are but histories, and they are "events" primarily only inasmuch as they were the *then* histories of the then events. But for the history of Greek mathematics, the works of Euclid, Archimedes, and Apollonius are themselves also subject matter and events, and not just histories of such, although they are this too. Newton's *Principia* is most certainly an "event," and, if it

[83]

is history, this history is history-then-in-the-making, namely the most "authentic" text for Newton's own theory, and for Newton's specific role in science. Thucydides' own theory of power politics, as known through his work, is to a considerable extent separable from the history of political events which his work narrates. Similarly, Herodotus' views on the role of the jealousy of the gods in the course of human events is separable from the actual human events which he narrates. But for Newton's *Principia* no analogous separation suggests itself.

The status of the *Principia* as a direct event, rather than as mere source material, is enhanced by the presumption that the work is also a work in natural philosophy proper. For instance, Newton's assertion, "hypotheses non fingo" —which in my view is much less consequential than sometimes assumed—has been searchingly interpreted with the same lexicological seriousness and finesse with which exegetes interpret the Bible or classicists interpret poetry and Presocratic fragments. Yet, in spite of the eminence of the *Principia* as a book by which to know Newton, it is less inappropriate to attempt to understand the spirit of Newton's theory by reading his book in translation, or even in paraphrase or epitome, than it would be to try to grasp the spirit of belles lettres by reading suchlike "substitutes" instead of the originals, or to try to grasp the teaching of Christ's Apostles by reading paraphrases of the Gospels, or perhaps even by reading them not in the original Greek. Or, to take another case, consider for instance the case of works of an Archimedes, say. Each such work, when in original, is a monument. But when it has only survived in a translation, or even only in an epitome, what has survived nonetheless becomes part of his collected works. It is true that a surviving translation, or epitome, has a value even in the case of poetry; but in the latter case the value lies more in what it suggests of the structure of the lost original than in what it is by itself. But

[84]

in the case of a work of an Archimedes, the content of the surviving substitute is, prima facie, of considerable interest for what it is by itself.

4. DOCUMENTATION OF GREEK SCIENCE

In the history of Greek science, incongruities in the distribution of the extant source material are exasperating. Generations of classicists have performed extraordinary feats of historical reconstruction, which are inexhaustibly fascinating to read about. But, of necessity, and admittedly, many reconstructions are fragilely founded, and some are, regrettably, nearly fictitious. In the Mediterranean and Mesopotamian areas there were intellectually controlled pursuits of science long before Greeks began to participate in them. But the beginnings of Greek participation are of extreme interest because Greek modes of scientific rationalization, whatever their origin, became, as such, the prototypes for later schemata. Now, the period of adolescence of Greek scientific rationality falls into the 6th and 5th centuries B.C., but any item of documentation for this lengthy period is either fragmentary, elusive, or indirect, and there are virtually no present-day specific statements about developments in this period which are, even *grosso modo,* beyond justifiable dispute or doubt. In science other than mathematics and astronomy, only the 4th century offers documentations of size; these are the works of Plato and Aristotle, in the main. Even in the case of these works there are disputes and doubts, but at least one need not dispute what the disputing is about. However, soon after Aristotle the picture blurs, and it never clears for the remaining eight centuries or so of the Hellenistic era. In mathematics, there is the glorious 3rd century B.C., with the works of Euclid, Archimedes, and Apollonius. But the preceding stages of the rise of this mathematics, during the three centuries or so before Euclid's *Elements,* are tantalizingly uncertain. At the other end, as already stated in the Introduction, after Apollonius darkness falls on the landscape of mathematics. There are only occasional streaks of illumination for the remaining six or seven centuries of it, and nothing can be told, other than in clichés, about the stages by which the gradual decline of Greek mathematics took place. Yet the knowledge of the morphology of the decline of Greek mathematics would be even more enlightening than the knowledge of the morphology of its rise. This would be so because Greek mathematics, by its very nature, was the first mathematics to have something of the texture of universality and of non-dependence on accidents of ambience, which are the overriding characteristics of the mathematics of today. And if Greek mathematics was nevertheless carried along into decay by the general decline of the Hellenism that was its ambience, then the particulars of this decay must have been quite powerful, and they must have been particulars not of accidents but of essentials; information about such essentials would be very informative indeed.

In Greek astronomy the first mathematically conceived system was that of Eudoxus (370 B.C.), and its impact on his contemporaries was such that the cosmology of Plato's *Timaeus* and the "theology" of Aristotle's *Metaphysica,* Book XII, did both concern themselves with it. Regrettably, there is no primary documentation about it, nor even a description in Ptolemy's *Almagest* (A.D. 150), and in the 19th century the system was reconstructed from much later sources.[47]

A methodological problem of some gravity arises when one interprets Greek natural philosophy as science proper. In modern science, that is, in science which has come into being since 1600, a string of asseverative statements counts for little unless the statements are connected with each other by an intellectually controlled nexus of reasoning.

[47] For references see Thomas Heath, *Aristarchus of Samos* (Oxford, 1913). The modern reconstruction is due to G. Schiaparelli.

Also, the "technical" language of science is expected to be relatively simple; a style whose comprehension presupposes the unraveling of lexicological subtilities is certainly not "scientific." However, much of Presocratic natural philosophy (and even of later periods) does not conform to such expectations, and one may doubt whether it is indeed science.

A major case in point is that of Heraclitus of Ephesus (500 B.C.), whom even antiquity christened the "obscure." [48] It is questionable whether it is meaningful to force his numerous sayings (*logoi*) into any kind of "theory," physical, cosmical, or other; in addition to being obscure from affectation, the sayings have the texture of aphorisms which were pronounced individually and in separateness of each other, and, Burnet notwithstanding, [49] the sayings of Heraclitus have indeed many structural similarities to the aphorisms of F. W. Nietzsche (1844– 1900), whom nobody would view as a scientist. Hermann Diels, the compiler of *Fragmente der Vorsokratiker* (many editions) refused to "attempt to arrange the fragments according to subject." This chagrined Burnet and others, but I dare say that Diels had an adequate understanding for the texture of scientificality of Presocratic natural philosophy, [50] and his refusal to attribute any kind of "system" to Heraclitus was well taken. Furthermore, it would not be sensible to affirm that the peculiarities of the style of Heraclitus are all due to the earliness of the science of his times. First, as already stated, Heraclitus was notoriously

[48] A recent book on early Greek philosophers including Heraclitus is W. K. C. Guthrie, *A History of Greek Philosophy,* I (Cambridge, 1962).

[49] John Burnet, *Early Greek Philosophy,* 4th edn. (1930), p. 132, footnote 5.

[50] The competence of Diels in assessing Greek physics is also attested to by the quality of his article (in *Sitzungsberichte der Akademie der Wissenschaften zu Berlin,* 1843, pp. 101ff.) about Straton of Lampsacus, pupil of Theophrastus.

"obscure" as if by intent. And, secondly, even Anaximenes of Miletus (546 B.C.) *before him* "writes simply and unaffectedly in the Ionic dialect" [51] and altogether gives the impression of having already been a "regular" physicist, similar to a "professional" of today.

Thales himself (585 B.C.), notwithstanding his extraordinary renown, is a somewhat shadowy figure. Aristotle was functioning as a "historian" when he tried to make a "system" of whatever sayings of Thales were known to him, but his "reconstruction" of Thales may have been more speculative than pertinent.[52]

Anaximander of Miletus, who was apparently somewhat younger than Thales, is very interesting in our present context. Plato does not mention him, and Aristotle does so only sparingly. There is no original utterance of his, although Theophrastus, the successor to Aristotle, apparently incorporated part of a sentence of Anaximander's into a larger sentence of his own. Apparently even Theophrastus, who was a man of considerable knowledge and savoir-faire, had difficulties in making him out. Nevertheless, secondary documentation is unanimous in suggesting that Anaximander was a person to reckon with, a kind of Alexander von Humboldt of his age. Anaximander's great claim to fame is the creation of the "philosophical" conception of *apeiron,* which was some kind of stuff (perhaps so tenuous as to be little more than "spatiality") of which the universe is made or composed. It is not known what this *apeiron* was, and speculations abound. The urge to speculate is irresistible, and yet I venture to assert, if only for the sake of argument, that it is not the task of history of science proper to do such speculating, not because no amount of speculation will

[51] Diogenes Laertius, Book II, Chapter 3; Loeb edition, I, 133.
[52] H. Cherniss, "The Characteristics and Effects of Presocratic Philosophy," *Journal of the History of Ideas,* Vol. 13 (1951), pp. 319–345.

[88]

restore Anaximander's own definition or description of it, if he ever had any, but because there is nothing to suggest that he had some kind of theory, that is, a nexus of interlocking statements, through which the "true" meaning of his *apeiron* was articulated and substantiated. Theophrastus, at any rate, did not transmit such a theory. And when trying to retain the flavor of Anaximander's own words by apparently weaving, as we have said above, a part of an original sentence of Anaximander's into his own, Theophrastus even apologizes for the quaintness (that is, "unscientific" manner) of the metaphoric style of Anaximander by adding that Anaximander was wont to express himself "poetically." It therefore seems doubtful to me, in this case at any rate, whether 20th-century lexicology, however advanced, can secure what Theophrastus already felt insecure of. An analogous situation would occur if in the year A.D. 4000, long after us, a historian of science would attempt to elucidate the scientific implication of the word "displacement" in Clerk Maxwell's very technical notion of "displacement current" by investigating the occurrences of the word "displacement" and its cognates in the realm of English Literature from *Piers Plowman* to Robert Browning. And one cannot dismiss this analogy on the ground that the Presocratics were still very sensitive to nuances of language whereas a contemporary scientist is not, because, as it happens, Maxwell was also quite attentive to idiom in all his writings, and there is no reason for asserting that Anaximander, or Parmenides, or Heraclitus was more so.

It is true that long before the stirrings of Presocratic philosophy proper, Homeric and later poetry had been extremely responsive to subtleties of linguistic differentiation. But it is most perilous to apply, as a matter of course, the knowledge of these subtleties in poetry to an interpretation of the philosophy afterward, just as it would be perilous to analyze, in detail, the novel scientific conceptions of the 17th century from a "professional" knowledge of the *Faerie Queene,* and of *Hamlet,* and of the works of the Pléiade, all from the 16th century. Also, in general Italian literature, Galileo Galilei is a founder of the Italian prose style, but no "literary" analysis could determine what Galileo "really" meant by "moments" in mechanics, because it meant to him, ambivalently, both our "momentum" and our "force." [53]

On the other hand I wish to observe that there are occurrences in Greek natural philosophy that are likely to be more disturbing to a humanist of today than an analogous situation in physics would be to a physicist of today. Anaxagoras (first half of the 5th century B.C.) was predominantly interested in the problem of the physical constitution of matter, and a number of fragments of his, in original, testify to his efforts. He introduced the notion of Homeomereity—the name itself may be due to Aristotle —and some philosophers are alarmed by ostensible inconsistencies in the characterization of the notion in different fragments. The late F. M. Cornford goes so far as to say that "unless Anaxagoras was extremely muddleheaded he cannot have propounded a theory which simply *consists* of this contradiction." [54] I think that Cornford's entire approach is wrong. As we have noted above, Galileo had, from our retrospect, a flat inconsistency in his notion of "moments," but nobody denounces him as "extremely muddleheaded" on that account. Also, Plato's dialogues are known in their pristine unabridged fullness, and not only in a few disjointed fragments as the book of Anaxagoras, or not even in patchwork versions, as some works of Aristotle. Yet a good part of the work of a Plato

[53] See to this E. Jouquet, *Lectures de mécanique,* Première partie (1908), pp. 24ff.; and Max Jammer, *Concepts of Force* (1957), p. 98.
[54] For references to Cornford and proposed solutions of the contradiction see Kirk and Raven (n. 9), pp. 367ff.

commentator consists in making sure that various passages from the same dialogue are indeed truly consistent with each other. But, quite apart from all this, in a later chapter [55] we will argue, with illustrations, that to science, in its ungovernable multiformness, inconsistencies are, *de facto,* much less of a bugaboo than to philosophy; and for the present we wish to adduce the following illustration. Between 1780 and 1840, in the pioneering thermodynamic work of Lavoisier and Laplace, Sadi Carnot, É. Clapeyron, and others, there were, simultaneously and quite knowingly, two separate, parallel conceptions of Heat, namely heat as molecular motion and heat as caloric fluid. There was a half-awareness that the two conceptions were not operationally equivalent; among other things, the operational employment of "caloric" sometimes, though not always, anticipated not the later conception of heat but rather the conception of "entropy." [56] Ernest Mach, author of the famed *History of Mechanics,* was, as philosopher, a positivist, and most positivists of his generation were "automatically" avowers of formal consistency. Yet when commenting on our above "illustration," Mach aptly observed as follows:

One learns from the history of thermodynamics that the intuitive notions, by which the comprehension of facts is facilitated and transmitted, somehow are of much lesser importance than the *precise* study of the facts themselves, by which the notions become adjusted and developed to the point at which their inherent constructive force becomes productive.[57]

[55] Chapter 4, sections 2 and 6.

[56] See Sadi Carnot, *Reflections on the Motive Power of Fire,* and other papers, edited by E. Mendoza (New York, Dover Publications, 1960).

[57] Ernest Mach, *Die Prinzipien der Wärmelehre,* 4th edn. (Leipzig, 1923), pp. 215–216.

Or, in original:

> Man lernt aus der Geschichte der Thermodynamik dass die veranschaulichenden Vorstellungen, durch welche man sich die Auffassung der Thatsachen erleichtert und vermittelt, doch eine viel geringere Wichtigkeit haben als das *genaue* Studium der Thatsachen selbst, durch welches eben erstere Vorstellungen sich so weit anpassen und entwickeln, dass dieselben erst ausgiebige constructive Kraft gewinnen.

That is, in physics the true meaning and scope of conceptions fully emerge only from the facts of their applications and not yet adequately from the imagery of their descriptions. This situation in physics has not changed at all since Mach but, in a sense, has in fact become more accentuated, as evidenced by the following quotation from Wigner's Nobel address:

> Equally remarkable is the present application of invariance principles in quantum electrodynamics. This is not a consistent theory—in fact, not a theory in the proper sense because its equations are in contradiction to each other. However, these contradictions can be resolved with reasonable uniqueness by postulating that the conclusions conform to the theory of relativity.[58]

In Greek mathematics, the great works from the 3rd century B.C. offer a panorama of its state at its height and in a period of finality, but reveal little, directly, about the stages of its rise since 600 B.C. A good part of the presumed knowledge about the earlier stages of mathematics has to be elicited from Plato, Aristotle, and suchlike sources, and this has somehow created in the minds of

[58] E. P. Wigner, "Events, Laws of Nature, and Invariance Principles," *Science*, Vol. 145 (1964), pp. 995–999; esp. p. 997.

historians a presumption, a most ill-founded one, that in the case of leading Greek mathematicians, their mathematical "ideology" is linked person by person to a philosophical posture of each. Thus, Agassi [59] takes it for granted, as did many others before him, that Euclid and Archimedes were Platonists of some kind and that the comprehension of their individual mathematical achievements is contingent on realizing this. It may be stated, however, that the mathematical texture of the works of Archimedes, and even of Euclid, is compatible with the presumption that their intellectual propulsion was purely mathematical, and mathematical only. Especially in the case of Archimedes there is, to me, nothing to suggest that he even knew, or cared to know, who Plato had been, or that he had more than a nodding acquaintance with philosophy and philosophers, if that much. It is true that the Stoic philosophers Posidonius (135–51 B.C.) and Geminus were competent in mathematics and astronomy. But one must not be led by this into assuming that Archimedes, before them, was for his part interested in philosophy, as if in reciprocation. One may well assert that Greek mathematics as a whole was an integral part of Greek philosophy, but this applies only to the overall mystique, not to details and specifics. Euclid, at the head of Book I of the *Elements* tried to observe a "philosophical" difference between "postulates" and "common notions," and in this he may have been following some philosophical trend of his day; but this does not in the least suggest to me that he was actively interested in the philosophical problem involved. A young instructor in physics nowadays is most likely to begin a course in mechanics by telling his audience that space and time are two a priori notions which cannot be further analyzed. But it would of course be incongruent to conclude from this

[59] See n. 1, p. 51 and p. 110.

that he is an adherent of Kantian philosophy and to try to determine the shading of neo-Kantianism to which he actually subscribes, although it is of course undoubtedly true that Kant's apriorization of space and time has had a lasting effect from which no physicist today can withdraw himself, even if he does not know that a Kant ever existed. We also note that Archimedes disregarded the would-be distinction between "postulates" and "common notions," [60] but that his indifference had, very posthumously, its philosophical rewards nonetheless. When D. Hilbert (1862–1943) turned his attention to axiomatics, first in mathematics and then also in mathematical logic, he first of all did away permanently with any distinction between various types of axioms in geometry; and he also taught logicians to be very sparing with such distinctions altogether, and his lesson has sunk in very deeply.

It is fair to say that by the time of Euclid mathematics was academically well separated from general and natural philosophy, probably in effect more so than it is in the 20th century. An intimation of such a separation may already be found in the dialogue *Theaetetus* of Plato. In the setting of the dialogue, Theaetetus appears to be a highly capable "graduate student" who has "majored" in mathematics under Theodorus of Cyrene, and in the dialogue Socrates examines him in philosophy which appears to be his "minor" topic. Theodorus is *ex officio* present, but he expressly declares himself academically not qualified to take active part in this installment of the examination (Plato 146 B), and his participation is indeed only nominal. Another mark of separation of mathematics from

---

[60] See n. 7, (i), p. 117–124; K. von Fritz, "Die ARCHAI in der griechischen Mathematik," *Archiv für Begriffsgeschichte* (Bonn), Vol. 1 (1955), pp. 13–103.

Archimedes, in the text as we have it, is even careless to the extent of not verbally distinguishing between axioms and definitions. See T. L. Heath, *The Works of Archimedes* (Cambridge, 1897), p. 2.

philosophy in antiquity may be seen in the attitude of the ancients towards the arguments ("paradoxes") of Zeno of Elea. Since the 17th century, and progressively ever more so, these arguments have been viewed as pertaining to the general area of foundations of mathematics rather than only to the area of philosophy of whatever kind. But there is absolutely no evidence from antiquity that professional mathematicians were aware of Eleatism, let alone were concerned with it. Even Plato in his dialogue *Parmenides* speaks of Zeno mainly as a disciple to Parmenides who is intent on protecting his master from abuse by opponents of the Parmenidean brand of monism,[61] but, except for a single fleeting allusion perhaps,[62] there is no awareness in Plato of the kind of Zenonian puzzle which Aristotle, with great foresight, has transmitted to posterity, to the benefit of both philosophy and natural philosophy.

Furthermore, apart from the fact that Plato himself, and a few intimates of his, liked to talk "philosophically" about mathematics, there is no evidence that this was the general tone in the Academy, or that the professional mathematicians among the academicians believed Plato to be the creative mathematician he liked to appear. Eudoxus was a powerful mathematician, and he was quite capable of presenting Plato with a planetary system which was all based on spheres, if that was what Plato wanted. But there is nothing to suggest that Eudoxus shared Plato's most trivial "philosophical" view that spheres are "divinely" or "transcendentally" beautiful.

And, as we will discuss more fully later on,[63] there is very little substance to the presumption that in the 5th century B.C., after the discovery of incommensurable

[61] For a history of preoccupations with Zeno from Plato till A.D. 1915 see the article in nine parts by Florian Cajori, "The History of Zeno's Arguments on Motion," *American Mathematical Monthly,* Vol. 22 (1915).

[62] *Phaedrus* 261 D.

[63] In Chapter 3.

magnitudes, Greek mathematics and natural philosophy went through a so-called foundation crisis.[64] Even if there really is an allusion to the "squaring of the circle" in line 1005 of the *Birds* of Aristophanes (first performed during the "theatre season" at Athens in 414 B.C.), this only shows that even Athenian playwrights "reacted" to "newspaper headlines," even ephemeral ones.

5. LIMITATIONS OF SUBJECT MATTER

The tendency in some areas of history of science to make much of little documentation has a "human" justification. The areas in question lie in a direction of a certain general appeal, but in this direction the subject matter is less abundant than sometimes assumed; in fact, in this direction the subject matter of history of science has even a tendency to shrink in time.

For instance, next to mathematics and astronomy, one of the oldest and most assiduously cultivated areas of systematic scientific thinking is the field of optics. Already in Aristotle it is taken for granted that there is an area of knowledge whose name is "optics." [65] From Hellenistic antiquity there are treatises about it by Euclid, Heron, and Ptolemy,[66] and in the Middle Ages the subject matter of optics was very much in the minds of philosophers, naturalists, and scholars in general, Arabic and Western.[67] Yet if one approaches the topic from a certain direction—

[64] The word "Grundlagenkrise" in German, which may have been coined by Herman Weyl, sounds even more fearsome than "foundation crisis" in English.

[65] For references see T. L. Heath, *Mathematics in Aristotle* (1949), especially pp. 11–12.

[66] See M. R. Cohen and I. E. Drabkin, *A Source Book in Greek Science* (1948), pp. 257–286. Also, O. Neugebauer, *The Exact Sciences in Antiquity* (1957), p. 226.

[67] See A. C. Crombie, *Robert Grosseteste, and the Origins of Experimental science* (1953); C. B. Boyer, *The Rainbow from Myth to Mathematics* (1959); V. Ronchi, *The Science of Vision* (1957).

which, although one-sided, has its rationale—one can also say that in a "true" sense the discipline of optics, as a genuinely systematic scientific pursuit, began only in the 17th century, first with the radically new theory of vision in Kepler's commentary on Witelo,[68] and secondly and principally with the sine-law of refraction of Snell and Descartes. This law was only an overture to a vast development which ensued, but it has the distinction that every textbook, on any level, makes mention of it more or less in the form in which it was first stated. There are not in physics many basic laws of this kind; some others are the two laws of Archimedes (law of levers and law of floating bodies), the law of falling bodies of (Oresme and) Galileo, the Kepler–Newton laws of planetary motion, the Huygens law on the elastic collision of mass units (which contemporary physics applies, without qualms, both to galaxies viewed as units and to elementary particles), perhaps Gilbert's law on the orientation of the magnetic needle, the Cavendish –Coulomb law of electric attraction, and the many-authored law $pv = RT$ of ideal gases.

I most definitely did not intend to say that there was no optics of interest before Kepler. On the contrary, the thoughts and intuitions in the optical works of Ptolemy, Alhazen, Grosseteste, and Witelo [69] are most intruiging to enlarge upon. But in a certain sense these works play the role of forerunners to what the 17th century actually initiated, and one must not shy from the statement—let alone denounce it as "naïve" and "unhistorical"—that all the developments in optics during the two millennia from 400 B.C. to A.D. 1600 were but an anticipation of what was achieved in the one hundred years of the 17th century by Kepler, Snell and Descartes, Olaf Romer, Erasmus Bartholinus, Huygens, and Newton. After this accumulation of achievements in the 17th century came a prolonged pause

[68] See V. Ronchi, preceding note, 40ff.
[69] See n. 67.

in the 18th century, which cannot be "explained" except by observing that 18th-century physics was concentrating largely on all parts of mechanics for discrete systems and continuous media. At any rate, in the beginning of the 19th century, as of a sudden, virtually all accumulated results from optics, not only in their original extent but with considerable augmentations were integrated by A. Fresnel (1788–1827) into nearly their present form; and all this during a mere stretch of 12 years of his all-too-short career.

The case of optics is not exceptional. A similar case, in which the historical complexity is greater and the historical lesson to be drawn even more emphatic, is that of mathematical logic. In the first decades of the present century, when mathematical logic began to be an academically organized discipline of its own, some adherents of the discipline began the twin task of reappraising the traditional logic in terms of the new one and of searching for antecedents of the new logic in the various compartments of the traditional one. In the course of these efforts, Medievalists began to form the view that the logicians of the Middle Ages had achieved more than generally assumed, in that they had done noticeably more than only contribute secondary modifications and additions to the corpus of Aristotelian logic with which, by a common presumption, they had been solely concerned. Historians of Hellenism went further, and they demanded even greater attention for their findings. They maintain, with evidential support, that in Stoic–Megaric circles, activity in logic went decidedly outside and beyond the doctrine of Aristotle, and that these circles arrived at conceptions which are characteristic of the propositional calculus of today. In fact, it is maintained, with justice, that the Stoics virtually introduced, and knowingly too, the twin notions of "truth-table" and "material implication." Because of this, J. M.

Bochenski, a leading historian of logic,[70] imputes to the Stoics an originality of the highest, and his affirmation of the achievements of the Stoics is so intense that he cannot but be censorious of anybody who thinks, or has thought, differently. Bochenski is particularly scornful of the voluminous 19th-century history of logic of Carl Prantl, which embodies a great anthology of excerpts from original works,[71] because Prantl's judgment of the Stoics, and also of medieval Schoolmen, is less than flattering. The anguish of some affirmers of Stoic and scholastic logic is compounded by the fact that they are also adherents of Immanuel Kant and therefore have to contend with Kant's cruel statement (1769) that logic since Aristotle had been unable to advance a single step and, to all appearances, had reached its completion; [72] although for Kant they somehow manage to find excuses in the end.

To all this I would like to observe as follows. As to bare facts, the critics of Prantl (and Kant) are right; by no means can this be denied flatly, as a defender of Prantl has recently attempted.[73] But the severity of the condemnation of Prantl is not justified, if in historical judgments one assesses achievements not only by their intrinsic level, as measured in isolation, but also by their impact on, and meaningfulness to, other developments which followed

[70] J. M. Bochenski is the author of several books; especially of *A History of Formal Logic,* translated from the German and edited by Ivor Thomas (1962), in particular p. 7.

[71] Carl Prantl, *Geschichte der Logik im Abendlande,* 4 vols. (1855–1870; reissued 1927).

[72] Bochenski (n. 70), p. 6.

[73] Günther Jacoby, *Die Ansprüche der Logistiker auf die Logik und ihre Geschichtsschreibung* (Stuttgart, 1952).

A temperate full-scale appraisal of the relation between the logics of Aristotelians and of Stoics is given in the work of W. Kneale and M. Kneale, *The Development of Logic* (Oxford, 1962), esp. Chap. III, pp. 113–176. This work offers an interesting, historically oriented corrective to the widely differing appraisals of Prantl and of Bochenski.

afterwards. Now, there is no evidence at all that Stoic logic had any significant impact on the evolution of our rational thinking of today; but nobody will dispute that the impact of Aristotle's own logic, in its pristine form or, say, in the variants of Ockham was immeasurable. Galileo, Descartes, Newton, etc. all fully knew Aristotle's *Organon* and were impregnated with its doctrines; but the leading Stoic logician, Chrysippus of Soli (280–204 B.C.), whom present-day historians of logic exalt, was probably unknown to them, even by name. Stoic logic may have had some indirect effect on the rise of Western thought, through works of Arabs into which it had entered, say, but if this was so, then the beneficiaries of the effect were certainly not aware of it. Also, Prantl belonged to the generation of many-sided scholars of status who came after the pioneering "philologists" of the Romantic and post-Romantic eras, whose knowledge was as wide as it was fruitful; and the mere fact that he had not become aware of the elements of originality in Stoic and scholastic logic is in itself a proof, albeit an inverted one, that within the stream of developments the role of these logics must have been a circumscribed one. Bochenski also complains [74] that Max Pohlenz, in his standard book on the Stoa,[75] devotes to Stoic logic no more than a dozen pages. But this complaint is self-answering. The aim of the work of Pohlenz is to assess the influence of Stoicism on the rise of Western civilization, and I think that within such an assessment Pohlenz gives to Stoic logic its fair share, perhaps not more than is its due, but also not less.

In sum, one ought not to reproach a succession of historians, from Prantl to Pohlenz, of not having discovered a situation of the past, without examining the question, which is equally historical, how it happened that

[74] See n. 70.

[75] Max Pohlenz, *Die Stoa, Geschichte einer Geistesbewegung,* 2nd edn. (1959).

they did not discover it, that is, without examining the question whether there really was as much to discover as one thinks there was.

From a certain pragmatic approach, all logic in the West was Aristotelian logic up to the time of Augustus de Morgan (1806–1878) and George Boole (1815–1864),[76] even though (Descartes and) Leibniz had already made a considerable effort to lead logic from "traditionalism" to "symbolism." [77] Some admirers of Leibniz and of mathematical logic bemoan the fact that in the 18th century contemporaries of Leibniz and others after them did not take the cue from Leibniz but allowed the origination of mathematical logic to wait till well into the 19th century. To this I wish to say, from hindsight, that, as developments went, everything turned out as well as one could wish for. The 18th century was extremely wise to give priority to constructing a thick basic layer of mathematics and erecting an edifice of rational mechanics; the 19th century was then the readier to initiate a mathematization of physics and of logic and the theoretization of other science. If this kind of rationalization of mine is too crass a case of "being wise after the event," then I wish to observe that, for my part, I have never felt dismay over savoring history, any history, backward through time, in addition to viewing it forward through time.

6. THE QUEST FOR SPECTACULARITY

In 19th-century physics one of the outstanding experimental achievements was the discovery of electromagnetic

[76] The principal work of Boole is *An Investigation of the Laws of Thought, on which are founded the Mathematical Theories of Logic and Probabilities* (London, 1854); new edition by Philip E. B. Jourdain (London, 1916). See also the refreshing book of Heinrich Scholz, *Concise History of Logic*, translated from the German by Kurt F. Leidecker (New York, Philosophical Library, 1961).

[77] L. Couturat, *La logique de Leibniz, d'après des documents inedits* (1907; reprinted 1961). See also F. Enriques, *Evolution de la logique* (1926).

waves by Heinrich Hertz. From being an indispensable phenomenon on earth and in the atmosphere, radio waves suddenly leapt into the prominence of also being the universe's medium of "communication" on an intergalactic scale. Nobody can foretell what this ceaseless intergalactic "chatter" via radio waves may still reveal. Yet histories of science do not expostulate much about it. They are much happier to dwell on the Michelson experiments about light velocity, which were nearly simultaneous with those of Hertz, and they simply delight to tell, retell, and re-retell the experimental findings of Ampère, Oersted, and Faraday from the first half of the 19th century. There are reasons for such preferences, some good, some precarious. The good reasons are that the Michelson experiments fall into the topic of "Time and Space," which seems philosophically irresistible, and that experiments like those of Ampère, Oersted, and Faraday have a fascination of immediacy and directness which simply cannot be argued against. But there is also a precarious reason, namely, that the experiments of Hertz do not conform, as much as do the other experiments, to a "popular" notion that a scientific achievement has to be a good-and-proper "discovery" in the sense of producing something that is unexpected and wonderously unforseeable in an obvious sense, and that if it is an experimental discovery, it ought to be independent from, and even run counter to, any preconceived theories and thus really "stump the other experts." What makes such an outlook precarious is the overwhelming fact that large parts of science are moving fast and inexorably toward a state of theoretization and mathematization in which this kind of naïvely thrilling discovery will become ever rarer and may become extinct altogether. For instance, the aim of Hertz was not at all to upset anybody's theory but, on the contrary, to verify one, and not even a theory of his own, but a preconceived theory of Maxwell, and one still to be really domiciled. Hertz achieved what he set out to do, gloriously so; but it was nothing "unexpected," and history of science does not quite know how to glory in his achievement.

Or take the case of chemistry. There seems to be an unceasing stream of historical studies on chemistry of the 18th and early 19th centuries. The presystematic pre-Lavoisier phlogiston phase of chemistry is receiving an increasing measure of attention, and some aficionados are asking for ever more of the same. But there seems to be no major historical study as yet on the unfolding of the twin theory of chemical valence and chemical bond, which has been provoking the originality and taxing the ingenuity of leading chemists since the middle of the 19th century, and that of atomic physicists since the beginning of the present century.

One might be tempted to retort to all this that, in physics and chemistry, from whatever reasons, the first decades of the 19th century are historiographically more alluring and challenging than the last decades. Perhaps on the whole they are. However, as we will outline in the Biographical Sketch of A. Fresnel (1788–1827), his case is, historiographically, quite similar to that of H. Hertz, although Fresnel was a great physicist in the very beginning of the 19th century. But the discovery of interference of light (that is, of the mutual enfeeblement of certain beams of light when superimposed) by his older contemporary Thomas Young (1773–1829) is again receiving the same kind of prolonged applause as do the various later discoveries of Ampère, Oersted, and Faraday.

In the theory of relativity, Minkowski's mathematical *mise en scène* was crucial and indispensable, and its elements of orginality are such that it is not at all certain that somebody else was bound to do it if he had not. But history of science does not have very much to say about this; just as it does not dwell upon the fact, emphatically, that the theory of relativity was a very organic continuation of the Maxwell–Hertz theory, whatever elements of "surprise" Albert Einstein may have injected into the context.

In philosophy of quantum theory, attention is riveted on the "spectacularity" of wave-particle duality with the connecting "uncertainty principle"; but no serious attention is given to the historical problem of finding out, in earnest, by which gradual and prolonged developments in the 19th century Planck's quantization of energy eventuated when it did. The account of Whittaker [78] is hardly final or even ample. It would be priggish, and futile, to inveigh against the enjoyment of the fascination of the uncertainty relation. But the cause of comprehending the course of physics would be served if some of the intellectual energies which go into the presentation of one more explanation of the true meaning of the uncertainty relation would be channelled into the construction of a comprehensive historical account of the rise of the notion of radiation of physical energy, beginning with Newton, say. Such an account ought to expound (i) how Planck's hypothesis in 1900 suddenly invested this notion with extraordinay status; (ii) how the notion of a "black body," which as a concept of physics already occurs in Newton,[79] was suddenly moved into the center of the stage of physics, although, as stated somewhere in a chance observation of Max Born, a black body is nothing but a humble, everyday oven; and, finally (iii) how, by inscrutable ways of man's creativity, Planck's originality was awakened only because Heinrich Hertz had constructed his vibrator ("resonator") in electrodynamics, so much so, that even 22 years after the event, in his Nobel address in 1922, Planck still expresses his indebtedness to the Hertzian constructiion.[80]

Next, in Hansen's important book on the discovery of the positron [81] very great stress is laid on the one fact that the experimental production of the positron by C. D.

[78] See n. 46.

[79] *Opticks,* Query 6; see also Lavoisier and De Laplace, *Mémoire sur la chaleur,* reprint (Paris, Gauthiers-Villers, 1929), p. 20.

[80] See Max Planck, *Collected Works,* III, 122.

[81] N. R. Hanson, *The Concept of the Positron; a Philosophical Analysis* (Cambridge, 1963).

Anderson in 1932 was achieved in a kind of half-ignorance of the great theoretical work, since 1928, by P. A. M. Dirac—and then soon also by others—on the quantum theory of the electron, by which the existence of a particle like the positron was strongly suggested. This one fact is certainly interesting and even significant, but I do not find that it is as pivotal and central as Hansen makes it out to be for the history of physics, or even for the appraisal of the stature of Anderson.

Also, in physics, the latest (before 1965) would-be spectacle which stumped the experts, and to which the public was also invited, was the discovery—that is, the corroborable assertion—of T. D. Lee and C. N. Yang in 1957, that certain elementary processes are not invariant under space inversion entirely by itself. The importance of this discovery continues undiminished, and the statement can be formulated with a minimum of mathematics. However, the true locale of the assertion is well inside our present-day physics as a mathematical system, and not really comprehensible by itself. Instead of becoming a spectacularity-in-permanence, this discovery quickly turned into an achievement-within-a-context; which seems to be the developing pattern for "surprise" achievements of the physics to come. Thus, the discovery of R. L. Mössbauer of 1958/59, which continues to have far-reaching consequences, is apparently viewed by most physicists as an initiation of a highly refined laboratory technique rather than as a bouleversement in which some cherished hypothesis or presumption was overthrown.

It may perhaps be said that indirectly the quest for spectacularity is receiving support from some of the leading "theories of theory," as, for instance, from the doctrine of K. R. Popper [82] that for the pertinency of a scientific theory its falsifiability is more important than its verifiability, and from the doctrine of T. S. Kuhn that

[82] See the work cited in n. 4.

major advances in science are normalizations of anomalies.[83] Nevertheless, it is my view that as the theoretization and mathematization of physics and of other science progress, there will be, all around, fewer and fewer discoveries of the old-fashioned spectacular and wondrous kind. The sooner history of science recognizes this and acts on the recognition, the easier will be its transition from adolescence to adulthood.

7. MEDIEVALISM

"Medievalism" is a certain impress, recognizable in its sameness, on strands of life and civilization, between Late Hellenism and the Renaissance, mainly in the mainstream of developments in the West, but also in tributaries from the East which are in confluence with it. In retrospect from today, medievalism was nowhere more pronounced than in science, and in no other history would it be more self-deluding to deny the occurrence of the phenomenon of medievalism than in history of science. In the development of Western science, medievalism is a kind of gap or hiatus or, at any rate, a kind of slowdown or muting of tempo. In the Middle Ages, developments in science did not always stand entirely still; in fact, in certain contexts or areas, scientific activity was quite lively, occasionally, on the face of it, even feverishly so. But, as appraised in retrospect, the outcome of medieval activity in science, even if not always small and transient, somehow did not articulate itself in works of living permanency or even of relivable actuality. Such was not the case in areas other than science. In medieval poetry there were heroic epics about Roland, Perceval, the Nibelungen, and Campeador, and others. There were sacred poems, immortal ones, like Dante's *Divina Commedia,* and "naughty" books like Chaucer's *Canterbury Tales* or the *Libro de buen amor* by Juan Ruiz

[83] T. S. Kuhn, *The Structure of Scientific Revolutions* (Chicago, 1962). For a "critique," see the following chapter.

(died, apparently, in 1351); and there were sonnets by Petrarch, whose appeal has been so enduring and whose style is so "modern" that he has been "postdated" into an "early" Renaissance man, as if "by decree." In architecture, medieval Gothic was an achievement beyond comparison with anything before or after. The grandiose efforts to harmonize faith and reason in works of Thomas Aquinas (1225–1274) and of others in the 13th century, which were preceded in the 12th century by works of Maimonides (1135–1204) [84] and others, have become an intimate part of our modes of rationalization in general. Also, the Middle Ages imbued "universities" for the promotion of knowledge with such a "universality" that even the nationalism of the Renaissance, with all its "Machiavelism," could not undo it. But in science proper there were no achievements at all which, by their level of attainment, would in any sense compare with the others. It cannot be asserted, without "explanatory" qualifications, that the *specific* modes of our "scientific" rationalization are a legacy from the Middle Ages.

In stating this, I most certainly do not wish to belittle the importance of unceasing and comprehensive studies of history of science of the Middle Ages. On the contrary, such studies are of first importance, and not enough can be had of them. It is a none too easy part of medieval history in general to assess the influence of medieval Arabism on the then European civilization; but in mathematics, in science, and in natural philosophy such influences can be more easily identified, intially at any rate, than in areas of humanistic and sociological attitudes. And the approach to the comprehension of the Middle Ages by way of science and natural philosophy is so important that, for instance, the effectiveness of the many-volumed history of civilizations of Arnold Toynbee for the appraisal of medievalism

[84] Within this context, the pertinent work of Maimonides in his *Guide of the Perplexed*.

is · severely circumscribed by the fact that the author's affinities to mathematics and science are minimal; the older work of Oswald Spengler, *The Decline of the West,* whatever its general shortcomings, has an advantage over Toynbee's work in this respect. But I do wish to say that, as it appears today, medieval science as a subject in history is less interesting to the general public than to the scholar, whereas science and natural philosophy of Greek antiquity have a rather wide appeal. Even Roger Bacon (1214–1294), who was an opponent of the philosophy of his times and a proponent of the cultivation of mathematics and physics, is by no means "popular" nowadays, in the sense of being read by the public at large. In his *Opus Maius,* Bacon has many passages which extol mathematics and science, but anthologies in history of mathematics and its applications do not as a rule include such passages.

In the history of science, medievalism is not only very pronounced, but it also begins earlier and terminates later than in other history, and its beginnings and endings are harsher, abrupter, and less easily motivated than in other history. Medievalism of science did not terminate, as other medievalisms, in the "Renaissance" of the 15th and 16th centuries, but only in the 17th century, in the setting of the so-called Scientific Revolution. In the 16th century, medieval Aristotelianism was still flourishing in Padua, and even in the first half of the 17th century Galileo[85] and Descartes were still battling furiously against it. Only in the second half of the 17th century does the battle come to a close, and, for instance, in Newton's *Principia* (1686) there is, finally, no discernible allusion to it left. Although the cosmological part of the Scientific Revolution reaches back well into the 16th century,[86] yet the first representa-

---

[85] See the illuminating comments in the edition by G. de Santillana of Galileo's *Dialogue on the Great World Systems* in the Salusbury translation (1953).

[86] The "revolutionary" work of Copernicus appeared in 1543.

tive of "modern" mechanics, Simon Stevin (1548–1620), appeared on the scientific stage rather abruptly, as if from nowhere, without any recognizable direct link either to the mechanics or the cosmology that preceded him,[87] in spite of the fact that Stevin frequently referred to F. Commandino (1509–1575) and perhaps even viewed himself as a "successor" of his. It is true that Stevin's near-contemporary Johannes Kepler did fuse mechanics with cosmology, but Kepler was such an unanalyzable compound of rationality and irrationality that no specific point in history can be proved or disproved by making reference to him. Furthermore, there are earnest judgments by *some* historians that the Renaissance itself, that is, the period between 1400 and 1540, say, made no substantive contributions to science proper.[88] But quite apart from such judgments, it is a fact that the subsequent rise of rational mechanics was, organically, by a discontinuity. Also, the early harbinger of the cosmological revolution, Nicholas of Cusa (1401–1464) seems to have appeared quite suddenly. Koyré, when assigning to him a predecessor, reaches back all the way to Lucretius (99–55 B.C.).[89] In general history, however, the transition from the Late Middle Ages to the Renaissance was such a gradual one that the very occurrence of the phenomenon "Renaissance" has been questioned.[90] But it would be very difficult to deny that

[87] It is a curious fact that the first great prophet in the historical books of the Old Testament, Elijah the Tishbite, also appeared as if from nowhere (I Kings 17:1).

[88] W. K. Ferguson, *The Renaissance in Historical Thought; Five Centuries of Interpretation* (Boston, 1948). This book presents the views both of those that affirm and of those that deny that the Renaissance was productive in science. To the affirmers also belongs Marie Boas, *The Scientific Renaissance, 1450–1650* (New York, 1962).

[89] A. Koyré, *From the Closed World to the Infinite Universe* (New York, 1958), pp. 5ff.

[90] Ferguson, see n. 87; see also Erwin Panofsky, *Renaissance and Renascence, in Western Art* (Copenhagen, Russak, 1960).

there was indeed a scientific revolution in the 17th century. Also the general questions as to what constitutes a revolution in science and which such revolutions there were in the past constitute an important theme in the history and philosophy of science,[91] whereas the corresponding question as to what constitutes a "renaissance" in general history hardly exists as a "theoretical" problem by itself.

At the beginnings of medievalism there are similar differences. In general history the earliest date for the onset of medievalism is A.D. 300. However, the representative medieval problem of harmonizing faith and reason is already outlined in Philo Judaeus (A.D. 40);[92] and in history of science there is a temptation to begin even earlier. Thus, after the great works of Euclid (320 B.C.), Archimedes (250 B.C.), and Apollonius (200 B.C.), the "classical" Greek mathematics suddenly and unaccountably lost its impulse, and in a certain sense the *Géometrie* of Descartes (1637) resumes where Apollonius and the commentator Pappus (end of the 3rd century A.D.) left off. Furthermore, the great *Almagest* of Ptolemy, which is Greece's legacy in astronomy and which Copernicus knew by heart, was composed around A.D. 150, nearly five centuries after Euclid; yet recent editors and commentators of the *Almagest*,[93] highly competent ones, have fully annotated it—in a very natural manner and not at all as a tour de force—by making references only to Euclid, with

---

[91] See n. 83.

[92] H. A. Wolfson, *Philo; Foundations of Religious Philosophy in Judaism, Christianity, and Islam* (Cambridge, 1947).

[93] Definitive text by L. Heiberg, 1899–1907; and most valuable is a German translation (with notes) by P. Manitius, 1912–1913, both in the Teuber Series of Greek texts. There is a "popular" edition with French translation in 2 volumes by N. Halma, 1813–1816, reprinted 192–; also a very readable English translation (with some annotations) by R. Catesby Taliaferro in *Great Books of the Western Worlds* (Chicago, Encyclopedia Britannica, 1952), Vol. XVI.

perhaps some references to Archimedes added. In order to appraise the enormity of the possibility of such an annotation, one ought to consider that an analogous situation would present iself if a great encyclopedic work in mathematical astronomy from the first half of the 20th century could be fully annotated mathematically, 1,500 years from today, by quoting only the *Algebra* of Rafael Bombelli (1572) with, say, the *Geométrie* of Descartes added. Even long after Ptolemy, the *Elements* of Euclid, with some parts of Archimedes, continued to be a sufficient prerequisite for astronomical works, and only Kepler began to go beyond that. But it is an outstanding fact that even Kepler's innovation, which was the one that brought Newton's *Principia* into existence, was, mathematically, not yet a proper "modernism" on the face of it. Kepler did not utilize some mathematics that later became a hallmark of his times; all he did, ostensibly, was to exploit the mathematics of the Greeks even more fully and fruitfully than his predecessors had done. Like his predecessors in astronomy he availed himself of the work of Euclid and Archimedes, except that, going beyond his predecessors, he also utilized, crucially, the work of Apollonius, which for 1,800 years had been lying around idly.[94]

One of the great perennial problems of history, any history, is to find reasons for the decline, that is the "dying out," of Hellenism. Now, no history, of whatever kind, can really "explain" this decline, but in general history there are at least full and informative treatises about it, such as Gibbon's *Decline and Fall,* the works of Rostovtzeff,[95] and in a certain sense even the *City of God* of St. Augustine. A book like that of Pirenne [96] which asserts that the Roman

[94] We will remake this point in Chapter 5.

[95] See M. Rostovtzeff, *The Social and Economic History of the Hellenistic World* (Oxford University Press, 1941), especially with emphasis on the "Summary and Epilogue" in II, 1027–1312.

[96] H. Pirenne, *Mohammed and Charlemagne* (New York, Norton, 1939).

Empire did not "really" cease to exist in the 5th century, but only in the 7th century, is most enjoyable and profitable to read, whether it be convincing or not. But, in the history of science, it does not seem possible to expound in full and enjoyable treatises how and why Greek mathematics was fated to decline gradually and then die out entirely in its own phase; or why the engineering feats of Heron of Alexandria and of his school did not initiate a trend in civilization but remained "toys" and "trifles"; [97] or why the Stoic attempts to build a physics [98] and a mathematical logic [99] left behind such a weak impress that the sharp-eyed scholars of the Renaissance and of the Baroque did not become aware of them; or why, for instance, it took *Homo sapiens* thousands of years to adapt the horse's harness to its body muscles. A very illuminating summary of various technological, political, socio-economic, and other reasons for the deterioration of Hellenistic science is presented in Neugebauer's book; [100] but the summary is only a challenge, and the problem remains.

The difficulty of explaining in full the decline of Greek mathematics is matched by the difficulty of explaining, in full, the rise of so-called algebra in the early and late Renaissance. This algebra, in its operational, that is, creative, aspects, is nowhere found in the known mathematics of the great Archimedes, and even what corre-

[97] Even Marxist-oriented works like that of Benjamin Farrington, *Greek Science, its Meaning for Us,* 2 vols. (Penguin Books, 1949), which submerge this question into a larger context, cannot spread the answer to it into a larger book. A recent *History of Ancient Rome* by the Soviet author N. A. Mashkin, 1947 (German translation *Römische Geschichte,* Berlin, Volkseigner Verlag, 1953) does not even deal with the particular topic of the decline of *science* at all.

[98] S. Sambursky, *Physics of the Stoics* (London, Routledge and Paul, 1959).

[99] See Pohlenz (n. 75), above; Benson Mates, *Stoic Logic* (California University Press, 1951); Antoinette Virieux-Reymond, *La logique et l'epistemologie des Stoiciens* (Paris, 195–).

[100] See n. 11.

sponds to this algebra in general modes of thinking is absent from Greek modes of knowledge generally.[101] By outward developments, this algebra came to the West from Arabic mathematics.[102] It entered Europe by way of Italy, from where it spread into France, Germany, and finally England, and it is the pride of "Italian" history of mathematics. But a leading Italian historian of mathematics, the late Gino Loria (1862–1954), would not allow himself to be drawn into formulating a historical judgment as to how and why it grew, flourished, and propagated itself as it did. As stated earlier,[103] as far as is known to me the only rationalization ever attempted is a socio-economic one. It may be strange, and even painful, to contemplate that our present-day mathematics, which is beginning to control both the minutest distances between elementary particles and the intergalactic vastness of the universe, owes its origination to countinghouse needs of "money changers" of Lombardy and the Levant. But, regrettably, I do not know by what arguments to disagree, when economic determinists, from the right, from the center, and from the left, all in strange unison argee.

8. PROGRESS AND CONTINUITY

After the Renaissance, and until the first decades of the 19th century, it was widely taken for granted that the

[101] See the last section in Chapter 1.

[102] How, in its turn, Arabic algebra came into being is a problem by itself, and a very difficult one too. It is even possible that its important ingredients came from a mixture of late Babylonian mathematics with an "underground" strain of late Hellenistic mathematics; if this was so, then the introduction of algebra into Europe in the Middle Ages was a reimportation of a kind.

[103] In Chapter 1, section 8. In addition to W. Sombart and W. K. Ferguson there quoted, we may mention Dirk J. Struik (*A Concise History of Mathematics,* New York, Dover, 1948, 2 volumes in one), and also Henri Pirenne. In the case of Pirenne it is not convenient to adduce any one specific passage from his writings on the Middle Ages and the Renaissance, but the general thesis is somehow there.

Middle Ages were intellectually "dark" ages, and that in the area of science this darkness was at its densest.

Now with regard to science there is an alternate view which, although in a sense even more dogmatic, has a certain air of plausibility which attracts adherents even in the 20th century; it is the view that "significant" science came into being only after 1600 or thereabouts, and that whatever in science happened before that date was only "preliminaries." This view may be opposed and rejected, but it cannot be scoffed at as being "naïve" and "unhistorical," because, for science at any rate, the date 1600 was indeed a dividing date as no other date, and not only on account of the fact that after 1600 science began to grow as never before. The decisive fact is that before 1600 science was territorial and that after 1600 it began to be universal, and very rapidly so. Before 1600 there had been an Egyptian mathematics, a Babylonian mathematics, a Greek mathematics, an Indian mathematics, a Chinese mathematics, etc.; a Greek physics, an Arabic physics, etc.; a Western medicine, an Eastern medicine, an African medicine, etc. But after 1600 there began to unfold, except for transient regional variations, a single physics, a single astronomy, a single mathematics, a single medicine, a single technology, etc. If underneath our present-day outward one-ness of the world there are separable centers of civilization nonetheless, then science, contemporary science, as a whole or in its parts, is least suited to offer criteria of separation that would be meaningful in depth. Also, in the light of this ever strengthening universalization of science since 1600, the ordering of all of science, even before 1600, only by chronology, as one science, as envisaged and done by Sarton for science since Hesiod,[104] is in effect less "naïve" than sometimes assumed.

Furthermore, the historical judgment that science truly

[104] George Sarton, *Introduction to the History of Science,* 3 vols. in 5 (1927–1948).

began only after 1600 can be made highly sophisticated if one introduces the concept of a "relative" beginning of a science, relative to a certain problem situation, that is. For instance, in the case of mathematics one may speak of a beginning of "anthropological" mathematics in prehistory when, according to a worn-out cliché, man began to cut notches and fashion tallies by which to count his sheep and establish his primitive calendar. Then there were the beginnings of a "systematic" mathematics in Egypt and of an "even more systematic" one in Babylon. Then came Greek mathematics, which started from an entirely new beginning, inasmuch as it was the first mathematics which was aware of a distinctiveness of itself, gave itself a name, and began to philosophize about itself. There was even, in the West, a medieval mathematics, with a beginning of its own, inasmuch as the work of Leonardo da Pisa (1200) burst in on its ambience as if from nowhere, unaccountably but firmly, and set the tone for almost four centuries. And parallel with these mathematics there were also "oriental" ones, Hindu, Arabic, Chinese, and Japanese mathematics, of which the first two strongly interacted with the "Western" ones. And finally after 1600 came our present-day mathematics, whose size and significance has been swelling uninterruptedly since. Whatever its similarities and links to the various mathematics of before, in effects and effectiveness this mathematics is an entirely new one, and the radicalness of this novelty cannot be glossed over by saying that this mathematics simply corresponds to a different phase in the everflowing system of altering or alternating successions of civilizations; or by making other such generalities, however intriguing they are.

In the case of physics the situation is even much more clear-cut. It is possible to maintain, not at all naïvely and almost irrefutably, that physics "truly" began only around 1600. In fact, from all of physics before S. Stevin (1548–1620), the only "basic" achievements which have textbook status nowadays are the two laws of Archimedes (lever, and floating bodies), and Archimedes and his contemporaries would have considered even these laws to be mathematics, anyway, although Aristotle had already distinguished between mathematics and physics, and also mechanics. Antiquity was well aware of Archimedes's prowess in devising various mechanical engines and was awed by it; but it nonetheless deemed him to be a mathematician, first and foremost.

Within the comprehensive process of "undarkening" the Middle Ages—which was set in motion in the first half of the 19th century and has almost become a distinguishing mark of historical scholarship in the 20th century—as far as science is concerned the most conspicuous single act of "undarkening," still under continuing analysis, is an assertion that certain findings of Schoolmen in the 14th century were anticipations of the geometry of coordinates and of the law of falling bodies, both from the 17th century.[105] There are unresolved differences of views about the extent of these anticipations and about their direct influences on 17th-century thinking, through public knowledge or "academic" media; and the stoutest advocate of the cause of the Schoolmen is Pierre Duhem.

Some parts of the very voluminous historical work of Duhem are diffusely descriptive and rhetorical rather than searchingly analytical. But, characteristically, his work is, intentionally or not, a large-scale undertaking to bridge the medieval gap in science which we have described in the preceding section on "Medievalism." As stated at the beginning of this chapter, Duhem's methodological approach was prominently continuist. Agassi [106] has aptly

[105] Leading research is due to M. Curtze, Pierre Duhem, H. Wieleitner, Anneliese Maier, E. J. Dijksterhuis, A. Koyré, Marshall Clagett, E. E. Moody, and others. For reference see the work of Clagett, *The Science of Mechanics in the Middle Ages* (Madison, Wisconsin, 1959).

[106] See n. 1, pp. 31ff.

stated that in Duhem's general conception there is in science always and continuously a certain amount of progression in evidence. The progression need not necessarily always be forward-directed by outcome, that is, an actual "advancement" in the accepted sense, but it manifests itself in a certain continuing and uninterrupted preoccupation with the subject matter of science, as a whole and in various divisions of it. In fact, Duhem, in his encyclopedic work,[107] actually forges what seems to be an unbroken chain of human links, from Thales to Galileo, clear across the entire Middle Ages without omitting a single decade or even a single year of them. He does not, naïvely, whiten out all the darknesses of the Middle Ages; but to Duhem the darknesses only indicate a certain lowering of the level of intellectuality by indentations in the flooring underneath, and not at all some chasmal rupture in the substance of the flooring.

But, as already stated, Duhem did not "create" continuism single-handed. On the contrary, continuism is fully operative in Aristotle and, I think, is due to him. There are stirrings of it in Plato, as when Plato suggests, albeit facetiously,[108] that Heracleitean flux had been anticipated by Homer, Hesiod, and Orpheus. Aristotle echos these suggestions of Plato unfacetiously; and, as if in response to these "cues" in Plato and Aristotle, beginning with the first half of the 19th century, some leading treatises and monographs on Greek natural philosophy have laid very-broad-gauge tracks from Thales all the way back to Homer. But in the case of Plato, apart from this kind of casual remark, it would be difficult to document expressly that he was, say, aware of an Ionic "school" of philosophy in the 6th century B.C.; for all that is known from the

---

[107] *Etudes sur Léonard de Vinci*, 3 vols. (Paris, 1900–1913); *Système du monde, Histoires des doctrines cosmologiques de Platon à Copernic*, 10 vols. (Paris, 1913–1959).

[108] Cherniss, n. 52.

dialogues, Thales, Anaximander, Anaximenes, Xenophanes, and Heracleitus may have been to Plato separate
individuals, cerebrating from separate spontaneities, and
without any continuist concern for one another. Aristotle
however, in his *Physica,* outright and most expressly posits
the existence of an Ionian School of hylozoist monists,
consisting, at least, of Thales, Anaximander, and Anaximenes. And then, by a tour de force, from an admittedly
irresistible urge for continuism, he even links to them
Parmenides and other Eleatics, although, as he himself
emphatically states, there is an insurmountable distance
between the physically oriented hylozoist monism of the
Ionians, and the ontologically oriented "noetic" and dis-
"criminating" (*krineiv*) monism of the Eleatics.[109] Aristotle thus created History of Philosophy as an academic
discipline, and his introduction of biases and slants, which
are notorious, is a small price to pay for an achievement of
such a magnitude. Also, any systematic history is, *au fond,*
biased history, well-intentioned maxims from Thucydides
to Ranke notwithstanding.

Present-day interpretations of Presocratic natural philosophy are full of continuisms, some of which are fragilely
or badly founded. One might aver that present-day speculative continuism in Presocratic philosophy is necessitated
by the fragmentariness of source material. But somehow
this in itself does not sufficiently explain why Aristotle
acted in this way, and why many Platonists nowadays, in
continuist fashion, insist that anything Aristotle said was
either a flat imitation, in disguise, of something Plato had
said, or an attempt, a futile one, to say something in defiant
opposition to what Plato had said, and, obviously, irrefutably better. And about the total extent to which
continuism is the avowed procedure nowadays, the following quotation from a recent, important work on Presocratics is unabashedly self-revealing:

[109] Aristotle, *Physica,* Book I, especially Chapters 2, 3.

The system of Anaxagoras, like that of Empedocles before him, and that of Atomists after, is to a large extent a conscious reaction to the theories of his predecessors. It will be easiest therefore, to base our reconstruction of it on his reaction to Parmenides, Zeno and other Presocratics.[110]

This kind of continuism has severe limitations; it would hardly be possible to make a reconstruction of the three planetary laws by "basing it on the reaction" of Kepler to Ptolemy, Copernicus, and Tycho Brahe. However, if in history of science the urge for interpretative continuism is greater than in other history, then this is understandable; it is due not only to accidental shortages of source material —which, after all, occur in all kinds of history—but even more to the substantive "Limitation of Subject Matter" with which we dealt in section 5.

Also, continuism as a method for studying the Middle Ages was attempted long before Duhem, and, in fact, long before the 19th century. It could probably be found in 18th-century "progressivists" like Montesquieu, Voltaire, and Condorcet, however scornful they might have been of the Middle Ages; and, at any rate, it occurs expressly in the 17th century. Gooch in his work on the 19th century observes [111] that Charles Perrault (1628–1703), in the *Literary Quarrel of the Ancients and Moderns* (*Parallèles des anciens et des modernes*), opines that the interruption of intellectual achievements in the Middle Ages was only an apparent one, and that the stream of events was like a river which for a distance flows underground.

Also, it seems to me that Duhem was a pronounced continuist more from a conjunction of circumstances than from a "free will" choice as historian. In fact, Duhem was a

[110] Kirk and Raven (n. 9), p. 368.
[111] G. P. Gooch, *History and Historians in the 19th Century* (1913), p. 8.

physicist by profession, a classicist by general upbringing, a medievalist by sympathy, and, apparently, a devout Catholic by conviction. Now, if a person like that sets out to write a history on any topic from Thales to Galileo, than I do not see how it could possibly be other than continuist. Finally, some deviations in the results of Duhem and successors have resulted from deviations of method which have nothing to do with continuism. Duhem, in linking the 17th century to the 14th century, proceeds from Nicole Oresme (1323–1382) via Leonardo da Vinci and other Renaissance persons to Galileo and Descartes; whereas, for instance, A. Koyré starts out from Galileo and Descartes, and traces their recognizable antecedents as far back into the early Renaissance as he can.[112] This is reminiscent, to a degree, of Huizinga's reexamination of the Renaissance. Burckhardt [113] traversed the period in a forward direction by starting out from its early phases in "Humanism"; whereas Huizinga [114] takes his stance in the early Renaissance, and from there casts a wide, searching look into the late Middle Ages, out of which the Renaissance had come into being. In an evolutionary sense it took first a Duhem to produce later a Koyré, just as it took a Burckhardt to produce a Huizinga; and the works of Burckhardt and Huizinga continue to be indispensable, side by side. It takes history to explain history.

9. CONVENTIONALISM

The 19th century began to clearly realize, and to openly avow, that it is not always possible or feasible to distinguish, exclusively and unyieldingly, between variant theories, or between variant modes of laying the foundations of a science, but that one may also have to tolerate the

[112] A. Koyré (1892–1964), *Etudes galilléennes,* 3 vols. (also in one) (Paris, 1939).

[113] J. Burckhardt, *The Civilization of the Renaissance* (Phaidon Press, 1944).

[114] J. Huizinga, *The Waning of the Middle Ages* (1924).

simultaneity of alternate theories, either because no decision between them, as rivals, can or need be made, or even, heretically, because one of them though "less true," is, for certain purposes, the "simpler" or more convenient one. For instance, even the most impassioned believer in the special theory of relativity may shrink from advocating that all present-day textbooks in all of science and technology in which anything is presented in "old-fashioned" Newton–Maxwell coordinates, should be instantly withdrawn and replaced by others in which only and exclusively Einstein–Minkowski space-time coordinates are employed.

In the course of the 19th century the growth of such general insights produced a methodological approach to the philosophy and history of science which, towards the very end of the 19th century, was articulated and advocated by H. Poincaré and is generally termed "Conventionalism." H. Poincaré was first and foremost a mathematician, and it is fair to say that his interest in physics, and in science generally, no matter how genuine, and how congenial to scientists, was nevertheless a reflection of the fact that in the 19th century, by influence of rational mechanics, other divisions of physics were becoming mathematized too. There was, especially in the first half of the century among adherents of the so-called *Naturphilosophie,* a certain amount of philosophers' doubt about the wisdom of letting mathematics spread too far and too deeply; this produced, unwittingly, a certain amount of reticence towards all-out mathematization, perhaps even within parts of physics itself. But there was something unnatural and even obscurantist in these would-be "profoundly" philosophical antagonisms against mathematics; and on the whole, the trend toward mathematization could not be countered and it has been irresistibly effective since. Even social science was affected by it. Economics, until then a "moral" philosophy, was subtly, or not so subtly, transformed when Cournot composed a mathematically oriented monograph about it.[115] Psychology became quantitative, in part but durably so, when G. T. Fechner (1801–1887) in 1860 injected into it his psychophysical logarithmic law linking stimuli and sensations; [116] and the efforts of 19th-century biology to use mathematics to divers purposes are all fully presented in the *summa* of Thompson.[117] On the other hand, in the second half of the 19th century, after an "inexplicable" pause of about 2,200 years, mathematics resumed in earnest the reflections on its own foundations which had been halted after their beginnings, in Greece, in the 5th and 6th century B.C., and which neither Euler, nor Lagrange, nor even Gauss had again in proper focus. And, having resumed its introspection, mathematics arrived at the conclusion that even in mathematics there are alternate possibilities for laying its specific and metaphysical foundations, so that, in any foundation of mathematics, in whole or in part, the ultimate "first" elements must be posited by an intentional and deliberate act of choice, that is, by a "convention," as Poincaré and his followers preferred to view it. Also—and this was very important—Poincare in a sense extended his "conventionalism" from mathematics to scientific knowledge in general.

Aristotle, in the second book of his *Posterier Analytics,* arrives at the reasoned insight that, barring an infinite regress, any "exact" knowledge must begin somewhere, sometime, from some (posited) beginnings. Aristotle frequently reiterated his stand against infinite regresses, but he somehow did it in schoolmasterly repetitive fashion, as a censorious self-admonition, and nothing constructive or fruitful ensued from this insight. Somehow his view of the

---

[115] A. A. Cournot, *Recherches sur les principes mathématiques de la théorie des richesses* (1838). Cournot had some predecessors whom he cites.

[116] *Elemente der Psychophysik* (1860).

[117] D'Arcy Wentworth Thompson, *Growth and Form* (1917 and 1942).

nature and role of mathematics remained sterile, and for scientific cognition no properly conventionalist approach emerged. Yet there is a certain feature of conventionalism, in a mathematical aspect, which reaches back into Hellenistic or even Hellenic antiquity, in astronomy at any rate. From various statements in the surviving works of Theon of Smyrna (A.D. 100), and in the *Almagest* of Ptolemy (A.D. 150), and also in the commentaries of Simplicius 5th century A.D.), it follows that already the astronomer Hipparchus, and also soon after him the many-sided Stoic philosopher Posidonius, clearly knew that it may be possible to "save" the same set of astronomical phenomena (to "save" = to "explain" or to "acount for") by alternate mathematical "hypotheses," that is, by alternate mathematical "models." The chief pair of equivalent alternate objects then known were "eccentricities" and "epicycles" as orbits of bodies in the Solar System, and it was clearly known that, although two alternate explanations may be equivalent on grounds of mathematics, one of them may nevertheless be preferable on grounds of physics.[118] Ptolemy even makes the thrilling statement,[119] that (in astronomy, of course) one ought to strive to have the mathematical model as simple as possible, but that, as a matter of course, the model may mathematicallly be as recondite as circumstances demand. This statement of Ptolemy's, as far as it carries, is immaculate conventionalism; [120] it cannot be easily gainsaid, and most physicists would endorse it. Furthermore, it is even possible to elicit from Simplicius a suggestion that the first beginnings

---

[118] We refer to two works: (i) Pierre Duhem, *Sozein ta phainomena, essai sur la notion de la théorie physique de Platon à Galilee* (Paris, 1908); (ii) Jürgen Mittelstrass, *Die Rettung der Phänomene* (Berlin, 1962).

[119] *Almagest*, Book XIII, Chapter 2, last paragraph.

[120] It ought to be said that Pierre Duhem (see n. 118) speaks not of "conventionalism" but of "positivism"; and Duhem implies that he himself is a positivist of this kind.

of this kind of "conventionalist" insight reach back to astronomers in the Academy, when Plato was still around.

Yet, by the unfathomable ways of Providence, the
Greeks did not fruitfully extend this magnificent philosophical insight from (their) astronomy to science and
scientific knowledge in general; nor did the Middle Ages or
the Renaissance do so. It is sometimes asserted by
Platonists that this kind of failure of the Greeks was due to
the influence of Aristotle, who counteracted efforts of Plato
to lead physics and science toward mathematization. But
the same Platonists also usually take it for granted that
Plato was far superior to Aristotle, and their assertion
would thus lead to the distressing conclusion that, in this
all-important matter, the Greeks of the 4th century B.C.
and after, instead of following a strong leadership of Plato
surrendered to a debilitating one of Aristotle. Even should
this conclusion be inescapable, I would still like to state
that in certain contexts Aristotle made an effort to aid the
mathematization of knowledge when Plato did not. In the
dialogue *Parmenides,* when reporting on work of Zeno of
Elea, Plato dwells only on Zeno's arguments regarding One
and Many, which have no overt mathematical implications,
and which, as Greek achievements, are certainly not
outstanding in themselves. Aristotle however, and only he,
transmitted to posterity Zeno's "puzzling" arguments
(*logoi*) against motion, "Achilles," and others, which,
rightly or wrongly, have stirred the imagination of mathematicians and philosophers ever since. And in the 19th
century, the impact of these *logoi,* which Aristotle so
foresightedly transmitted, had a greater effect on the
creation of our present-day philosophy of mathematics
than all of Plato's own mathematical philosophemes taken
together.

From a certain approach, perhaps a one-sided one, one
may assert that Isaac Newton's creation of his theory of
gravitation was an achievement of conventionalism—the

greatest ever. In fact, mathematically Newton does nothing other than fashion a hypothesis for the saving of a certain aggregate of kinetic phenomena. However, the heart of the achievement was a unique and unprecedented manner of forming the composition of this aggregate. The total aggregate was the merger of two subaggregates, one constituting "celestial" phenomena, and the other "sublunary" ones. Aristotle in his book on the *Heavens* was already wondering, long and loud, how to yoke these two sets of phenomena. But nobody before Newton had the vision, or daring, or inspiration, or the sheer greatness, to fuse them as simply and directly as Newton did, and for ever too.

Even before H. Poincaré elevated his conventionalism into a doctrine in philosophy, it had been *de facto* encountered as a necessity in physics. Thus, Clerk Maxwell has a well-articulated conception of a scientific theory, and of a mechanical or mathematical model of a scientific theory; the idiom of his background language sounds like ours. One might perhaps interpose that in Maxwell a physical model, no matter how much constructed by himself, has a certain "absolute" reality nonetheless, which seems to contravene the main conventionalist tenet that the choice of models is free, and therefore their reality a "relative" one. To this I can retort as follows. Almost any theoretical physicist of today will, when in a mood for philosophizing, avow some version of conventionalism. But during his "working hours" he nevertheless "believes" in an almost "absolute" reality of the "data" of physics, and even staunchly so. He "believes" in the reality of the weirdest of elementary particles, even if he has to admit, before man and God, that these particles are constructed by himself, qua physicist, out of nothing other than some notions and techniques of mathematics. And just because so much of physics is made of sheer mathematics, the physicist of today may even be led to "believe" in the "reality" of the objects of this mathematics with an intensity which exceeds that of a mathematician, who is not so "motivated."

10. THE MATHEMATIZATION OF SCIENCE

The unceasing systematization and mathematization of events in science introduces into its history a unidirectedness and irreversibility which is far greater and much more unrelenting than in any other history. We have already noted above, in section 6, that this forces upon the historian a subtle but significant reorientation of attitude towards spectacularity of achievement. A more general and comprehensive effect is this, that no past phase of development need ever structurally recur, not even in strong similarity, whatever ostensible analogies may suggest themselves. This produces the following historian's dilemma, which, although familiar from all history, is more poignant in history of science than anywhere else. On the one hand, the very fact that past phases cannot recur makes it doubly desirable that history faithfully reproduce a past phase, as it "really" was, within its own context and against its own background. But, due to the same process of progressive systematization, it happens, even frequently, that the original structure of a past phase becomes much more fully understood and intelligible through subsequent developments—much more so than in other history—so that, in a peculiar manner, the very aim of knowing events of the past, even as they were in the past, makes it not only desirable but also very pertinent and even mandatory to examine them through their continuation into later events and through their living link with the present.

For instance, it is nowadays very strenuous and even very difficult to really comprehend the entire physico-mathematical content of Newton's *Principia* (1686) in its full technical detail by studying this basic work exclusively in the mathematical imagery of Archimedes and Apollonius, in which Newton elected to fashion it. The difficulties are not due only to our present-day estrangement from this mathematical idiom of the Greek past; they are *intrinsic* difficulties which were inherent to the work itself from the beginning. By its idiom, the *Principia* is ostensibly backward-oriented. However, by nearly unanimous and very strong testimony of the mathematicians of the 18th century, by content, the *Principia* was very forward-oriented, decisively so, and was full of incipiencies, and full of only partially articulated anticipations of developments which were bound to be engendered by it. When the 18th century began to articulate these incipiencies and anticipations, the tendency toward doing this in another idiom, namely in the analytical idiom of our present-day textbooks, became irresistible almost from the beginning. And to ask a historian of today to persevere where the great mathematicians of the 18th century were not able to do so, whether from incapacity or from unsuitability, would be almost *unwissenschaftlich,* no matter what a Ranke of history of science might legislate.

Another aspect of the same difficulty manifests itself in the fact that if one proposes to analyze the genesis of the *Mécanique analytique* of Lagrange, which was composed in 1786, one hundred years after the *Principia,* and in an entirely different idiom of analysis, it is almost impossible to decide which of its data and consequences were due to Newton and which were definitely not so. For our comprehension, Newton's mathematical schema in the *Principia* has become both too elusive and too allusive. Even some of the beginning parts of the *Principia,* which are relatively easy to approach, may become nearly meaningless when taken *au pied de la lettre* and in isolation. It is my impression that interpretations of them by historians and philosophers are much more specific than ought to be permissible, the reason being that somehow Newton's modes of mathematico-physical cerebration are not, or not yet, as rationalized as are those in 18th century, say. For the same reason, the famous *scholia* of the *Principia* are certainly over-interpreted. The intellectual redaction of the *Principia* is such that the supremacy of the work rests in what Newton actually achieves and not in what he says in the *scholia* that he does.

A variant on this last difficulty occurs frequently when the authorship of an achievement is "hyphenated" without being "joint." For instance, the ruling physical theory of the 20th century is quantum theory, and its role in science is immeasurable. Its originator was Max Planck, and it grew out of a problem of G. R. Kirchhoff, the first answer to which had been the so-called Stefan–Boltzman radiation law. Now, when historians of science eventually attempt, as they should, to trace the antecedents of quantum theory with the same kind of fervor with which they investigate the antecedents of Newton's theory, they will probably wish to find out which of the two authors, J. Stefan or L. Boltzman, was "really" responsible for the law named after them. But this will not be easy to determine from the pertinent publications of the two. The publications suggest that they worked separately, even though they may have been personally acquainted. Stefan was only an experimentalist, and, within a purely experimentalist setting, he derived a certain formula purely from observations.[121] Afterward, apparently without making any independent experimentation of his own, Boltzman "verified" the same formula by actually deriving it from electromagnetic theory, and, within its setting, really convincingly.[122] One may speculate that without this theoretical derivation the formula might not have entered at all into the

[121] J. Stefan, "Beziehung zwischen Wärmestrahlung und Temperatur," *Sitzungberichte der Wiener Akademie,* Vol. 79 (1879), p. 391.

[122] L. Boltzman, "Ableitung des Stefanschen Gesetzes betreffend die Abhängigkeit der Warmestrahlung," *Annalen der Physik und Chemie* (Wiedemann), Vol. 22 (1884), p. 291.

awareness of physicists in general, and of Planck in particular.

Furthermore there are scientific situations which, when arising, are quite strange, but to which there are later parallels which are less strange; a later parallel may aid in allaying the strangeness of an earlier one. For instance, the great Presocratic philosopher Parmenides introduces, as his own creation, an "Existent" (*to on, to einai*) which, to begin with and for quite a while, seems to be a purely noetic construct, without apparent reference to physical materiality. But this changes gradually, and at a certain point, and then with jarring suddenness, the Existant becomes endowed with attributes of a (cosmic) homogeneous material sphere of finite radius.[123] Now, in my private efforts to accommodate myself to the thought patterns of Parmenides, I have derived little satisfaction from the numerous commentaries,[124] however brilliant they may be; but I have been aided by the following parallel. When, toward the end of the 19th century, electrodynamics finally posited the existence of an electron (the name is due to G. Johnston Stoney, 1891), the prime intent was to introduce a definite elementary quantity of electricity by itself, that is, only a particle (= minimal unit) of "electric charge" without any involvement with mass, inertia, or size. But the nexus out of which and into which the concept was constructed demanded what, at the time, most physicists also took for granted, that an electron also "have" an inertial (= Newtonian) mass, and "be" a small homoge-

[123] Fragment 8, lines 43–45, in Diels-Kranz, 6th edn.; John Burnet, *Early Greek Philosophy* (Meridian Books, 1962), p. 176, second paragraph.

[124] The latest book on Parmenides with very ample references is Mario Untersteiner, *Parmenide, testimonianze e frammenti* (Florence, 1958); to our present question see pp. clxiff., 150ff.

An even more recent, detailed study is the book of Leonardo Tarán, *Parmenides, A Text with Translation, Commentary, and Critical Essays* (Princeton University Press, 1965).

neous sphere of some radius. This entire imagery is as "primitive" as anything in the Parmenidean poem, but its role in contemporary physics is important and unshaken.[125] Undoubtedly every physicist wonders, every now and then, whether there might not "exist" an electric charge which is entirely divorced from mass, just as he apparently wonders, most of the time, why the amount of mass has to be what the so-called measurements say that it is, never more, and never less; but that "Stoney's" unit does have a mechanical mass is beyond question. Also, the very specific values of the Newtonian mass of the various elementary particles of today (A.D. 1965) are so oddly and unaccountably assorted, that any "wild" statement by a Presocratic is tame by comparison.

Unlike other events and their history, science and history of science are, at the same time, very old and quite young. By method, history of science has to observe, from general history, some proven rules of great stringency, and at the same time it also has to fashion rules of its own which only further developments of science can suggest and produce. And the history of science is the most uncertain of all histories, because in a sense all of science till today has been but a wink before the forthcoming awakening of science on the morrow.

[125] R. A. Millikan, *The Electron* (1917, 1924); also his article "Electron" in *Encyclopedia Britannica,* 14th edn. In Chapter 2 of the book there is also a full report on the naming of the electron by Stoney.

CHAPTER 3

REVOLUTIONS IN PHYSICS
AND CRISES IN MATHEMATICS *

1. THE THESIS OF T. S. KUHN

WE WILL now deal with two topics which, although separable, are closely connected with each other. The first and larger part of the chapter is concerned with the conception of a revolution in physics, as recently blueprinted in a provocative book by Thomas Kuhn.[1] We will make observations which are seemingly in conflict with those of Kuhn, but I really intend to amplify and qualify some of Kuhn's theses rather than to dissent from them, and my approach is somewhat different anyhow. After that, we will make some observations on revolutions in physics as far as the underlying mathematics is concerned. And, finally, we will make some remarks on so-called foundation crises in mathematics, which may be viewed as a kind of revolution, and especially on a major crisis of this kind which is presumed to have taken place in the 5th century B.C.

Kuhn, in his investigations into the nature of revolutions in science, analyzes both the inward ontological and epistemological nature of such revolutions and the psychological and behaviorist attitudes, resistances, and responses of practitioners of science, before, during, and after a revolution. Kuhn finds that revolutions in science are mostly internal revolutions, brought about by some scientists and then forced by the initiators on the scientific community at large. There is even an implied suggestion

* Originally in *Science,* Vol. 141, No. 3579 (2 August 1963) pp. 408–411. Copyright 1963 by the American Association for the Advancement of Science. Somewhat revised.

[1] T. S. Kuhn, *The Structure of Scientific Revolutions* (University of Chicago Press, 1962).

that, in the beginning, a revolutionary innovation may be both desired and resisted by the same group of scientists, ambivalently. Kuhn makes a point of emphasizing that most scientists all the time, and all scientists most of the time, prefer peace to revolution, normalcy to anomaly, and the preservation of their "paradigms" to changes of paradigms, a "paradigm," according to Kuhn, being more or less a sum of "universally recognized achievements that for a time provide model problems and solutions to a community of practitioners" (see p. x of the Introduction to his book).

This finding is indeed meaningful, and as already noted by Gillispie,[2] it is one that can be easily accepted. For my part, I found nothing singularly disturbing in the realization that among scientists, as in other groups of human beings, the revolutionaries of today are likely to be the conservatives of tomorrow; that paradigms are not readily abandoned or changed unless anomalies make it imperative; and that there may be diehards who will not give in even then. But if one *is* surprised and disturbed to find that resistance to innovation is widespread and even dominant among "professors" who are expected to be professionally pledged and conditioned to emphatically seek the truth and nothing but the truth, I think that there is no clear reason for singling out scientists from among scholars in general. Kuhn's diagnosis of innate conservatism does attach a certain stigma, and it is restrictive to the entire study to stigmatize scientists for something which philosophers and humanists also practice.

Perhaps we can see evidence of the humanists' concern to preserve a paradigm in a well-attested event [3] which

[2] C. C. Gillispie, *Science,* Vol. 138 (1962), p. 1251.

[3] The main reference is to an article by Otmar Schissel in *Real Encyclopädie der klassischen Altertums-Wissenschaft,* edited by Pauly-Wissowa (1930), Vol. 28, cols. 1759–1767; the article is an authoritative interpretative digest from R. Asmus, *Das Leben des Philosophen Isidoros, von Damaskios aus Damaskus* (Leipzig, Philosophische Bibliotek, 1911), Vol. 125.

occurred in the Neoplatonic school at Athens during the last 50 or 60 years before its dissolution by the Emperor Justinian in A.D. 529. At the time, leading circles in the school were opposed to the increase in Aristotelian features in the "official" world picture, the result of certain influences from within. A leading exponent of Aristotelian ideas was Marinus, born in Neapolis (the Hebrew Shechem) in Samaria, who eventually became head of the school, succeeding the much-adored "divine" Proclus (A.D. 411–484). Marinus, when still in the junior position of tutor, was in charge of Isidorus, a student. One day Marinus showed Isidorus a commentary on the platonic dialogue *Philebus,* which he had just composed. But, on the authority of Damascius, biographer of Isidorus, the latter prevailed upon Marinus to destroy the commentary, on the strange grounds that their great master Proclus, then head of the school, had already composed a commentary on the *Philebus* to end all such commentaries. Present-day scholarship maintains convincingly that the true motivation for this request was apprehension lest the work of Marinus inevitably show a bias against the Neoplatonic "paradigm" even if Marinus made an effort to keep it out.

2. THE ROLE OF MAX PLANCK

One of Thomas Kuhn's star witnesses is none other than the physicist Max Planck. In *Scientific Autobiography,* written in 1937,[4] Planck remarked that "a new scientific truth does not triumph by convincing its opponents and making them see the light, but rather because its opponents eventually die, and a new generation grows up that is familiar with it." This harsh statement has the ring of an undeniable verity, but it so happens that it hardly applies to

[4] See either M. K. E. L. Planck, *Physikalische Abhandlungen und Vorträge* [Collected Works] (Brunswick, Vieweg, 1958), III, 389; or M. K. E. L. Planck, *Scientific Autobiography and Other Papers,* translated by F. Gaynor (New York, Philosophical Library, 1949), pp. 33–34. Kuhn quotes the passage in *The Structure of Scientific Revolutions* (see n.1), p. 150.

Planck's own discovery. It is true that Planck's discovery was not hailed on the instant as the great breakthrough that it was, but it is equally true that Planck did not have to wait for anybody to die before a considerable measure of recognition was meted out to him.

Although Planck's first papers on quanta began to appear only in the latter half of 1900, as early as 1911 renowned physicists of the day convened at a Solvay Congress in Brussels with an agenda devoted entirely to the new theory, and with Planck as the honored speaker.[5] Before that, in an encyclopedic article [6] dated May 1909 and entitled "Theorie der Strahlung," the radiation specialist W. Wien, although unsympathetic to Planck's theory, gave it exhaustive coverage. On the other hand, Planck himself, as late as 1922, in his Nobel address, almost tried to play down the originality of his achievement by pointing out how his intense preoccupation with the then relatively new Hertz vibrator (or "resonator") was the catalytic setting that eventually brought forth the quantum hypothesis.[7] In fact, in the list of references to the written version of the address, Planck cites the paper of Heinrich Hertz in *Annalen der Physik* [8] in which Hertz discusses the vibrator which he had constructed for purposes of testing the Maxwell theory. With this vibrator Planck, in 1900, by a kind of "ideal" construction in the black body, devised his quantization. It is regrettable that,

[5] P. Langevin and M. de Broglie, *La theorie du rayonnement et les quanta* (Paris, Gauthier–Villars, 1912); A. Eucken, *Die Theorie der Strahlung und der Quanten* (Halle, Knapp, 1914).

Among the participants at this Congress were such luminaries as Madame Curie, A. Einstein, J. H. Jeans, P. Langevin, H. A. Lorentz, Kamerlingh Onnes, H. Poincaré, E. Rutherford, A. Sommerfeld, and J. D. Van der Waals.

[6] *Encyclopädie der mathematischen Wissenschaften* (1904–1926), Vol. V3.

[7] See M. K. E. L. Planck, *Physikalische Abhandlung und Vorträge* (see n. 4), III, 122.

[8] H. Hertz, *Annalen der Physik,* Vol. 36 (1889), p. 1.

in after years, Planck and his biographers (especially von Laue) rarely restated this illuminating biographical fact, if they ever mentioned it again.

In general, I would say that between 1900 and 1925, in the quantum theory "revolution" which was brought about by Planck himself and then by Niels Bohr and others, the changeover from one paradigm to another was peaceful and evolutionary, without any of the characteristics of a revolution by *bouleversement,* and without manifestations of a forced normalization of an anomaly, as Kuhn envisages it. One might retort that, although this particular transition was a peaceful one, yet the emergence of the theory of relativity within the same period, 1900 to 1925, and in fact within the narrower period 1905 to about 1918, did indeed conform to the Kuhn pattern, and that the second half of the quantum theory revolution, which began soon after 1925 and in which Heisenberg, Schrödinger, Dirac, and others were the protagonists, did also. To this I would reply that what gave the emergence of relativity the character and status of a revolution was not its effect on physics proper but its effect on the notions of space and time, which, long before that, had become objects of paramount importance to philosophy in general. We will try to show later in this book [9] that space and time did not become paramount in general philosophy until after the Renaissance, whereas in classical Greece, for instance, in spite of a strong trait of spatiality in the general imagery of rational thinking, space and time, as specific notions, were notions of physics and of physics only. If the concepts of space and time had not attained the preeminence in general philosophy which they had through developments that had occurred since 1600, relativity would have been much less of a revolution than it was.

Furthermore, the sudden outburst of interest in quantum

[9] In Chapters 4 and 5.

theory and in Planck's constant came after the proclamation of the uncertainty principle by Heisenberg, in 1925, and here again the interest arose from the fact that Heisenberg, in his "popularization" of his principle, emphasized the involvement of the notion of ordinary space. The over-all fact that the uncertainty principle applies to pairs of conjugate operators in general, if stated without emphasis on particular pairs which correspond to ordinary Newtonian coordinates of position and momentum, would hardly have caused philosophers the malaise they felt when the emphasis was placed on these coordinates.

Also, while the cosmological revolution in the 16th century did indeed replace one paradigm with another—namely, the geocentric with the heliocentric—it cannot be said that the theory of relativity replaced Newtonian space with other spaces, in the sense of making Newtonian space obsolete or antiquarian. Planetary and particle mechanics, vast stretches of "phenomenological" mechanics of continua, and much of statistical mechanics, of thermodynamics, and even of electrodynamics continues to be Newtonian; any serious attempt to de-Newtonize them would make most of mechanics, much of physics, and virtually all of engineering unrealistically complicated and would be widely resisted. And even within the heart of physics, which did indeed become genuinely relativistic, there is no single paradigm in control, such as a Kuhn revolution by *bouleversement* would have terminated in. In fact, the Lorentzian space of quantum field theory and the substratum space of most cosmological models now in vogue are totally contradictory. Thus, for instance, in the Lorentzian universe, space and time are inseparably fused; but in most other cosmological models there is, as of yore, a recognizably separate time axis, even if the artless, simplicity of this separateness is no more.

Finally, I wish to point out that, to many physicists, the theories of relativity, whatever their éclat, were terminal phases of the era of Newton, Lagrange, Hamilton, and Maxwell rather than initial phases of a new era. But Planck's original quantum hypothesis, even if its advent was rather peaceful and even if its antecedents in 19th-century radiation theory were comparatively unspectacular, was apparently the prologue to a tremendous epic-still-to-come, of which only the introductory scenes have been playing thus far.

Kuhn's notion of a scientific revolution may subsume too many possibilities under a single formula. The formula apparently is meant to apply to the emergence of modern science in the 17th and 18th centuries, viewed as one giant revolution, in spite of its size and in spite of the fact it was much more the direct emergence of something entirely novel than the transformation, by revolution, of something old into something new. But the formula is seemingly also meant to apply to many particular events, which occurred in succession but are viewed separately, such as the many turns and twists and even vagaries in the presystematic phases of electricity and chemistry in the 17th and 18th centuries, especially the 18th century. And finally, the formula might end up by being applied, through a circular mode of identification, to any kind of changeover which has a visible and recognizable trait of originality and creativity associated with it. Now, the question of what constitutes originality and creativity in *Home sapiens* is probably one of the most difficult problems in philosophy and philosophical sociology, and not much would be gained by reducing the problem of what is a revolution of knowledge to the problem of what constitutes creativity in man. Also, it is probably feasible to distinguish between latent and active phases within creativity. This would lead to a corresponding distinction, during non-revolutionary periods in science, between genuine "normalcy," and "latency" of anomaly, and the evaluation of this distinction, within Kuhn's schema, might become very troublesome indeed. Furthermore, in physics there seem to be periods of concentrated creativity which are made up of rapid successions of many small but sharply defined discontinuities of achievement, and to such periods Kuhn's formula of a "normalization of an anomaly" can be applied only with difficulty.

3. CRISES IN MATHEMATICS

A novel situation arises, in the case of physics, if we turn our attention away from physics proper and toward the scheme of mathematics that underlies it. This mathematics seems to have paradigms of its own, and they are more inward ones. These inward, structural paradigms seem to behave differently from the outward, purely physical ones, sometimes even disturbingly so. Thus, neither electricity nor magnetism had any mathematical paradigm at all until, toward the end of the 18th century, Cavendish and Coulomb formulated the so-called Coulomb law, and until, at the beginning of the 19th century, Poisson initiated magnetostatics and electrostatics by introducing a mathematical theory of potentials into the context. In this sense, in spite of Lavoisier, one must say that chemistry began in earnest only with the laws of Dalton, Avogadro, and Dulong and Petit, all in the beginning of the 19th century. I do not mean to suggest that there were no disciplines of electricity, magnetism, and chemistry before the 19th century. But I do mean to suggest that, to somebody in the field of mathematics, the cataloging of many separate revolutions which these disciplines are supposed to have gone through in the 17th and 18th centuries is bewildering and unconvincing.

A similar observation could even be made about the course of "classical" Greek physics in its entirety. The developments from Thales to Aristotle are frequently presented as a "motivated" succession of revolutions in which an emergent would-be physical or cosmological system was knowingly and militantly put forward to supersede an earlier system. But the fact is that Archimedes, who was the only mathematical physicist Greece ever produced, seemingly refused to be involved in these crazy-quilt developments, and to his sober mind they probably appeared to be "irrational" and unmotivated.

On the other hand, in the case of mechanics, physicists and other scientists are wont to view the 17th and 18th centuries as one unit of development. And yet, mathematically,[10] Newton's *Principia* has the appearance and many of the attributes of the works of Archimedes and Apollonius, whereas Lagrange's *Mécanique analytique* (1788) is not radically unlike textbooks of today, 20th-century modernisms notwithstanding. Also, the mathematico-physical subject matter of the various divisions of mechanics which took shape in the 18th century was not at all a mere explication of what had already been presented in the *Principia,* implicitly if not expressly, even though the effect of Newton's treatise on later developments was an overwhelming one. In fact, by "juggling" mathematical paradigms one could make out a case for the assertion that there was a much greater distance between the mechanics of Euler and Lagrange and the mechanics of Newton than there was, in the 19th century, between the electrodynamics of Maxwell and Hertz and the hydrodynamics of, say, Helmholtz.

Apparently the mathematization of a science affects the role and nature of revolutions that may and do occur in it. And since, on the other hand, most of science is tending toward mathematization, even determinedly so, one should guard against generalizing from the shape of premathematical revolutions to the shape of revolutions in general.

Mathematics itself also has its revolutions, and developments in the 20th century have led to the singling out

---

[10] See Chapter 6.

among them of revolutions of a particular kind, which are termed "foundation crises." It has been asserted, with emphasis and even with a dash of sensationalism, that the classical Greek mathematics which is known from the works of Euclid, Archimedes, and Apollonius went through such a crisis in an early Pythagorean phase in the 5th century B.C. The cogency of this assertion may be questioned, and the assertion has indeed been contested. But the clamor for a retroactive crisis in antiquity has been such that temperate counterassertions have not been able to mute it. An "anomaly" in Greek mathematics did indeed emerge; it was the discovery that the square root of 2 is irrational, or rather that, in a square, the diagonal is incommensurable with the side.[11] Greek mathematics did certainly react to this discovery, with attentiveness and resourcefulness as is evidenced by the reference to the problem in the Platonic dialogue *Theaetetus,* and by the erection of a Greek theory of incommensurables, apparently attributable to Eudoxus, which is the subject matter of the fifth book of Euclid. However there is no indication in Greek doxography of a "crisis" or of a "mathematical scandal," except perhaps for the late-Pythagorean tradition that Hippasus of Metapontum, an unruly, early member of the Pythagorean sect, was violating rules of the sect by divulging to outsiders details about research-in-progress on incommensurables and other problems, and that he suffered divine retribution for his indiscretions.[12] One might perhaps also adduce the fact that, in Archimedes's work on "Sphere and Cylinder I," in

[11] It has also been suggested that it was with respect to the regular pentagon, not the square, that the incommensurability of the diagonal and the side was first recognized (see K. von Fritz, *Annals of Mathematics,* Vol. 146 [1945], p. 242).

[12] For references relating to the "crisis," see Walter Burkert, *Weisheit und Wissenschaft, Studien zu Pythagoras, Philolaos und Platon* (Nurenberg, Carl, 1962), p. 432, and "Hippasos" in the index. For information about Hippasos in particular see also M. T. Cardini, *Pitagorici, testimonianze e frammenti* (Florence, "La Nuova Italia" Publishers, 1958).

the prefatory letter to Dositheus, the puzzling assertion that the mathematical procedure of Eudoxus is "most irrefragable" [13] indicates that even at the time there were some "philosophers" who were not satisfied with the manner in which Eudoxus resolved the crisis of incommensurables. Against these very slender items of support one has to note that there is no allusion to a mathematical crisis or "scandal" in any of the passages in Aristotle from which, with due caution, most of what is known about Pythagorean mathematics and principles of science has to be abstracted.

A systematic theorizing about foundation crises in mathematics was begun about 50 years ago in response to the challenging discovery, around 1900, that there are paradoxical situations in George Cantor's "naïve" set theory and hence in mathematics as such. In fact, the first foundation crisis was identified, in substance rather than in name, in 20th-century mathematics itself, and past crises were then uncovered in the wake of this one. The "Greek crisis" theory was received very attentively and sympathetically all around, perhaps in remembrance of Zeno's puzzles, which to some philosophers are inexhaustibly provocative. But I should point out that the authors of a recent book [14] are trying to arouse interest in a third

[13] T. L. Heath, *The Works of Archimedes* (Cambridge University Press, 1897), p. 2.

[14] A. A. Fraenkel and Y. Bar-Hillel, *Foundations of Set Theory* (Amsterdam, North-Holland, 1958), pp. 14–15.

It appears from an account in W. Kneale and M. Kneale, *The Development of Logic* (Oxford, 1962), pp. 652–653, that the alarmist outlook on foundation crises in mathematics was institutionalized in 1902 by Bertrand Russell via G. Frege, when Russell threw Frege into a panic by pointing out to him the foundational weakness ("set of all sets") of G. Cantor's naïve set theory, which Frege had been using freely. It was on the eve of the publication of Volume 2 of Frege's *Grundgesetze der Arithmetik, begriffschriftlich abgeleitet,* Volume 1 of which had been published in 1893. In a Postscript to Volume 2, Frege is abject and penitent, and he cannot do other than quote: *Solatium miseris, socios habuisse dolorum* (It is a consolation to a wretch, to have fellows in pain).

foundation crisis, and the "anomaly" which underlies this crisis is the one whose normalization consisted in the "rigorization" of analysis in the 19th century by Cauchy, Weierstrass, Cantor, and others. This last alleged crisis is the one least deserving of the name. Perhaps it was "anomalous" for the infinitesimal calculus to pile up achievement upon achievement in the 17th and 18th centuries, without being frustrated by inadequacies of mathematical rigor, and to become introspective as to its rigor only afterwards; if so, this was the healthiest and the most wonderful "anomaly" that could have occurred. Newton, the Bernoullis, Euler, d'Alembert, Lagrange, Laplace, and others made advances beyond anything one might have asked for.

To say that their achievements landed mathematics in a "crisis" is incongruous, unless one is prepared to aver that Thales and Pythagoras plunged rational thinking into a crisis-in-perpetuity, inasmuch as they introduced mathematics into rational thinking inseparably, and inasmuch as there is no prospect of constructing a logico-ontological foundation of mathematics that will be absolutely and unqualifiedly satisfactory to all, forever.

CHAPTER 4

ARISTOTLE'S PHYSICS AND
TODAY'S PHYSICS *

IN MAKING some observations on the nature of the physics of the Greeks, especially of the physics of Aristotle, our aim will be to point up similarities and dissimilarities, analogies and discrepancies, between the physics of antiquity and the physics of today. The aim is not to find out how much of Greek physics survives in ours, but to verify that to a certain extent the similarities in the modes of general thinking, ancient and modern, also reflect themselves in parallelisms between the doctrines of physics which these modes of thinking have produced.

1. PHYSICS AND MATHEMATICAL PHYSICS

Greek physics, while being observational and perhaps also experiential, never became properly experimental. The Greeks, of both the Hellenic and Hellenistic eras, never developed, or even truly initiated, a kind of physics that would correspond to the so-called theoretical or mathematical physics of today. This was not at all due to the fact that Greek physics was not "ready" for it. In fact, modern physics, that is, the physics which began to develop after the Renaissance, started out from rather modest beginnings, but it showed *from the very beginning* a tendency toward becoming mathematical, and toward becoming so in a manner and to an extent that is a hallmark of its "modernism" and an instrumentality for its ever growing success.

Nevertheless, as already stated in Chapter 1, the Greeks

* Originally in *International Philosophical Quarterly,* Vol. 4 (1964), pp. 217–244. Somewhat revised.

did have areas of knowledge in which science and mathematics interpenetrated. Their astronomy was clearly mathematical, from Eudoxus to Ptolemy. Also, as early as Aristotle there was, on his testimony, something called *optike* which was subordinate to geometry, and something called *harmonike* which was subordinate to arithmetic (and according to the *Posterior Analytics* 78 b 37 there was even a kind of *mechanike* that was subordinate to "stereometry").[1] Furthermore it follows from various known statements in Aristotle that Pythagoreans envisioned some kind of mathematization of physics,[2] although the extent and depth of their insights is not easy to appraise even when the allusions to them in Plato's *Timaeus* are added to express statements in Aristotle.[3] In a known passage in the *Physica* (Book II, Chapter 2, 193 b 22–194 a 15) Aristotle even seems to be wondering, for a fleeting moment, whether astronomy, *optike,* and *harmonike* can indeed be distinguished from mathematics proper. But in the same passage, and in passages in the *Metaphysica,* he has no such doubts about what he invariably and firmly calls *physike,* and which apparently corresponds to physics in general in our sense. He insists that mathematics and physics (that is, Greek mathematics and Greek physics) are distinct and separate, and that this separateness is not one that can be overcome. In this, as it turned out, he was right. This separateness was indeed never overcome by the Greeks, not even by Archimedes who could have done so, or by physicists and mathematicians after him who should have done so. His laws on the balancing of the lever and on floating bodies clearly pertain to mathematical physics and were the first of their kind. But they did not have the effect of initiating a mathematical

---

[1] For references see Thomas Heath, *Mathematics in Aristotle* (Oxford, 1949), especially pp. 11–12.

[2] *Ibid.* (see index under "Pythagoreans").

[3] See F. M. Cornford, *Plato's Cosmology* (New York, 1937).

physics at the time. Archimedes the mathematician was revered in antiquity, and Stoic and Epicurean philosophers who specialized in physics and cosmology must have been aware of him. However, there is nothing to indicate that any of these philosopher-physicists was capable of comprehending that Archimedes was a species of physicist, too, or that any of them had an inkling or premonition, if ever so vaguely, that it would be this species of physics to which a later future would belong.[4]

2. PHYSICS AND PHILOSOPHY

The Loeb edition of Aristotle's *Physica* by Wicksteed and Cornford, which is a very good commentary on the contents of the treatise, begins its "General Introduction" (p. xv) with the following paragraph:

> The title 'Physica' is misleading, and the reader must expect to find little or nothing that it suggests in this treatise. 'Lectures on Nature,' the alternative title found in editions of the Greek text, is more enlightening. But 'Principles of Natural Philosophy' (as the term would have been understood in the eighteenth and earlier nineteenth centuries) would be better still.

I do not agree with this statement, and I find that it is generally less misleading to view the *Physica* as a book on physics, pure and simple, than as a book on philosophy in whatever sense. In my own endeavors to make sense out of Aristotle, that is, to find coherence and meaning in his problems and in whatever answers he finds for his problems, I found it revealing and rewarding to approach Aristotle's treatise as a general book on physics, literally and ingenuously. Aristotle is frequently discursive, repetitive, and resumptive, and especially in the *Physica* he

[4] A. N. Whitehead, *The Concept of Nature* (Cambridge, 1930), p. 24.

sometimes involves himself in equivocations, inconsistencies, and even contradictions which discourage readers, distress and alarm commentators, and encourage detractors gratuitously. Now, philosophy has always been very sensitive to inconsistencies, and it is probably best that it should be so. Physics since the Renaissance has been professing to be so too, but only when in a holiday mood of solemn introspection; when in a workaday mood of extrovert activity it has been much less attentive to the pitfalls of inconsistencies, and sometimes it has even been outright oblivious of them. It so happens that in the last part of the 19th century some very articulate leading physicists, especially French ones (Pierre Duhem, also Henri Poincaré and others), were sternly positivistic, and were demanding that the philosophers' ban against inconsistencies also be strictly applied to the edifice of physics as an indivisible totality and at all times. But in the 20th century, developments in various parts of physics have been precipitous and uncontrollably multiform, and strict formal consistency for physics as a whole is not enforceable. Physicists, almost eschatologically, profess to wish and pray for consistency in the future, but they are willing to forego it "temporarily," at any time, for the then present.

In the light of this, if I view the *Physica* and related treatises of Aristotle as works on physics, and not on philosophy, then many commentators' difficulties in these works lose their poignancy, in that they appear to be much less damaging to the value and meaningfulness of the whole than commentators fear. In fact, if I allow myself the license of drawing up appropriate parallelisms to situations from contemporary physics, some difficulties in Aristotle, although not removed thereby, can be made to enhance the value of Aristotle's physics instead of subtracting from it. An instance of such a "saving" interpretation will be given later, in section 17, to a notorious difficulty in Aristotle's theory of *topos*. But before this other matters will be dealt with.

3. TIME IN PHYSICS

Aristotle has an essay on time in *Physica* IV, 10–14, and we will begin our comment with a quotation from the Loeb edition:

> As to Time, Aristotle enters no profound metaphysical speculation as to its essential nature. He is content with attempting to bring precision into the thoughts of the plain man, who may easily fall into confusions and contradictions when he tries to give himself an account of what he really means by it. (I, 378)

No part of this statement has any pertinence if one views the *Physica* as physics. Aristotle does not address himself to "the plain man" but to the physicist, and he does not indulge in metaphysical speculation because time is to him a concept of physics, and most exclusively so. He does mention "time" in other treatises, but only sparingly, and in no way does he analyze it outside the *Physica* in any detail. It is important to keep always in mind that to Aristotle time and space were only objects of physics and not also of his First Philosophy, as they did gradually become in the course of the 23 centuries since.

As viewed from present-day physics one can say that Aristotle is struggling to establish or clarify the following facts, in thought patterns of his own.

(1) Time is a "determinant" of any motion or change in physics, so that any physical process runs it course in time ("Time is the number of motion").

(2) Time in physics has to be represented by a certain mathematical schema, which makes it into a *"megethos"*

(magnitude), and qua *megethos* time is mathematically a so-called open linear continuum, which is also unidirected ("arrow of time").

(3) Although time is a unidirected open line, nevertheless the only way of measuring it is by cyclical, that is, periodic, events.

(4) The oneness of the universe is somehow connected with a oneness of time. (This is implied in *De Caelo,* I, 9)

(5) There is no awareness of time without a perceiving soul.

We note that (4) and (5), if viewed as problems, have been raised "independently" by physicists in the 20th century. Furthermore, statement (3) is a profound truth, and it originated in the *Timaeus*. We also remark that until the "discovery" of the pendulum in the 17th century no "reliable" cyclic event other than stellar motion was available, and that competent astronomers knew that only astronomically observed time was true time. (The time of 20th-century atomic and nuclear clocks is observed spectroscopically.)

4. PHYSICS AND COSMOLOGY

It is probably right to say that many of Aristotle's significant statements on time already occur in the *Timaeus,* expressly or implicitly, and that there is nothing in Aristotle to match Plato's stirring phrase, "Time is the moving image of eternity." We wish to observe, however, that Aristotle's statements on time in the *Physica* are physics and have pertinence in the physics of today, but that Plato's phrase would not have a specific meaning within physics or even physical cosmology of today. The *Timaeus* may be admired as a great book, but it is nonetheless a fact that nowadays for a compendium on physics and cosmology the mythologically cosmogonic setting of the *Timaeus* would be intolerably archaic and backward looking; whereas Aristotle's division of his work into a sober and matter-of-fact treatise on physics, which comes *first,* and a non-cosmogonical book on the heavens, which comes *afterward,* would be in keeping with arrangements of compendia today. Aristotle may even have been the first to introduce such an arrangement knowingly. Furthermore, Aristotle's total dispensing with "myths" may be viewed as a turning toward "scientificality," no matter how deeply Platonists may draw inspiration from myths in their master's discourses.

The *Timaeus,* after the introductory conversation, has a solemn invocation of the gods, and Plato then descends downward from a divine on-high to a cosmic ordering of the universe, and thence to the actual creation of it in detail by chance, necessity, and law. Aristotle, however, after the manner of a "regular" scientist, ascends upward. Such a scientist begins by dwelling fully in physics proper, after which he very likely will want to advance to problems of general cosmology, analytical or evolutionary. And after that he may even feel an urge to move on further to questions of ethics, morality, and theology.

In briefer words, to Plato physics was a particularization of cosmology, whereas to Aristotle cosmology was, as it is prevalently today, an extension of physics. This is a significant difference of orientation; it is neither "minor" nor only "technical," and it most certainly does not indicate a paucity of imaginativeness in Aristotle's whole make-up, as Platonists sometimes conclude. This difference of orientation affected Aristotle's approach to many problems, especially to the problem of space, to a significant extent, as we will see. But we will next discuss Plato's principle of necessity and Aristotle's reaction to it.

5. NECESSITY AND CHANCE

A penetrating insight into Plato's *ananke* (necessity), against the background of 19th-century knowledgeability, is the high point in George Grote's interpretation of the

Timaeus in his work on Plato. I like best the following single sentence:

> By necessity Plato means random, indeterminate, chaotic, preexistent spontaneity of movement or force: ἡ πλανομένη αἰτία upon which Reason works by persuasion up to a certain point, prevailing upon it to submit to some degree of fixity and regularity.[5]

Cornford's *Plato's Cosmology* has an elaborate analysis of necessity which is a rationalization of Grote's finding,[6] and Cornford quotes not the above sentence but the following passage:

> We ought here to note the sense in which Plato uses the word Necessity. This word is now usually understood as denoting what is fixed, permanent, unalterable, knowable beforehand. In the Platonic *Timaeus* it means the very reverse:—the indeterminate, the inconstant, the anomalous, that which can neither be understood nor predicted. It is Force, Movement, or Change, with the negative attribute of not being regular, or intelligible, or determined by any knowable antecedent or condition—Vis consili expers.

Cornford draws Book II of Aristotle's *Physica* into his analysis, and he rightly concludes that underlying its Chapter 8 there is a notion of "Necessity" or rather of the expression "of Necessity" (ἐξ ἀνάγκης) which closely conforms to the interpretation in Grote. To this we wish to add the following observation. In the preceding Chapters 4, 5, 6 of Book II, Aristotle has manifold reflections on the role of *tyche* (chance) and *Automaton* (fortuitous) in the relations between man and man and between man and nature.

[5] *Plato and the Other Companions of Socrates,* 3rd edn. (London, 1875), III, 266.

[6] Pp. 159–77. The quote is from Grote, p. 249.

If viewed in retrospect, one might say that in these reflections Aristotle was groping for an extension of Plato's *ananke* from the realm of the physical to the realm of the anthropical, that is, from the area of physical science to a would-be area of social science—for an extension of the physical science of Huygens and Newton to the social science of William Petty and James Bernoulli. The *Ars conjectandi* of Bernoulli (published posthumously 1713; he died 1705), which inaugurated so-called laws of large numbers in probability and statistics, is very much concerned with Grote's subordination of the "random and indeterminate" to the persuasion of mathematical reason, with the aim of securing a measure of "fixity and regularity." It extends its concern mainly to problems of human and moral standing of the kind that preoccupy Aristotle in Chapters 4, 5, 6 of Book II, and it is this which made Bernoulli's book a starting book for all of social science to come. It is worth noting that the corresponding subordination of the "random and indeterminate" in the realm of physical "spontaneity of movement or force" was begun only in the 19th century, with the work of Clerk Maxwell and L. Boltzmann. However, as if to compensate for the lateness of the beginning, in the 20th century this development entered deeply into, and also created, novel theories of physics which seem to be reaching out even more persuasively for a subjugation to our peculiar reason of the universe as a whole.

In order to avoid misunderstanding, let it be stated that we do not assert that Plato's *ananke* was the direct antecedent of some conceptualization in modern physical science, or that Aristotle's *automaton* was the direct antecedent of some conceptualizations in modern social science. Greek physics and even Greek mathematics had each run down its own course in late antiquity, and modern physics and even modern mathematics began from a new beginning. But something important in our manner of

[151]

rational thinking comes from the Greeks. They created it and now it is fully ours. Because of this there are many traits of similarity between Greek physics and our physics, and they force themselves on our attention. An injudicious exploitation of such similarities can be misleading, but it is nevertheless not necessary to dismiss them from our attention. When prudently assessed such similarities can be quite informative, both about the Greeks and about ourselves.

6. *Chora* ("SPACE") AND *Topos* ("PLACE")

In Plato, space (and time) occurs mainly in the *Timaeus*. Although it occurs in the dialectical part of the *Parmenides,* it is referred to but very sparingly in the ontological analyses of the *Theaetetus* and *Sophistes.* As for Aristotle, so also for Plato space is primarily a notion of physics, or at most of physical cosmology, but for Plato this does not hold quite so firmly as for Aristotle. The reason is not that for Plato space is some kind of "philosophical" entity, as for Kant, say, but in the *Timaeus* physics has not yet come out entirely from the cosmological "Receptacle of its Becoming." In fact, the very name *physike,* as also the names *mathematike, mechanike,* and even *poetike,* solidified only in Aristotle, and very quickly and durably so; they have hardly changed a shading of their meaning since. This in itself was an Aristotelian achievement which is rarely seen in its magnitude.

Plato and Aristotle have two appelations for spatiality, χῶρα and τόπος. *Chora* is frequently broader, and it is less specific and determined than *topos*. A locus in mathematics, that is, a figure which is determined by, or results from, specific requirements, became *topos* and not *chora*. In the *Meteorologica,* when Aristotle wishes to single out a geographic district in a country, *chora* usually stands for country and *topos* for district. Also, *chora* seems to be the room occupied by a general object, and both in *Timaeus* and *De*

Caelo, if "occupying" is indicated by a verb from the root ἔχειν, then, apparently for idiomatic reasons, space is *chora* as a rule.

Apart from that, in the *Timaeus* space in its cosmogonic nascency (52 A, 53 D) is *chora* and it continues to be so designated while being "shaken down" into shape (53 A, 57 C) and even after that for a while as long as it still may be empty = κενόν (58 A). But when finally firmly "emplaced" and when it has become "gravitational" space, in the sense that some bodies in it are bound to move from certain positions into other ones, then it is *topos* (57 C, 58 C, 60 C). In *De Caelo* Aristotle scrupulously observes the same distinction, but since his cosmology is less cosmogonic, the occurrence of *topos* prevails.

Aristotle also uses the name of *topos* for the conception of place which he creates in *Physica* IV, Chapters 1–5. It is an original creation of his, and in Chapter 2 he asserts with pride that it was not anticipated by anybody, not even Plato. Aristotle's conceptions of place would fit well into the physics of today, better than into the small amount of physics of his day he had to press them into. Viewed and interpreted in retrospect, what Aristotle was aiming at was a notion of spatiality that would serve as a certain setting for a general "physical system," thus being a properly conceived attribute or attributes of a substance or of a collection of bodies rather than mere "extension" in a naïve sense. Aristotle's great advance over his predecessors manifested itself in a clearly discernible aim—which however was marred by a serious miss—of creating a space of physics, as a datum of physics, and of physics (and science) only; whereas Plato created first—with the aid of receptacles, matrices, bastard reasonings, and what not—an all-comprehensive space of cosmology, and then obtained the "ordinary space" of physics and of "individual" physical events from the all-comprehensive space by a downward specialization only. What Aristotle was groping for was what is today a physicist's "space of the laboratory." If a physicist nowadays wishes to conceptualize the spatial setting of an indoor experimental design, then he ordinarily does not imitate the *Timaeus* and engage the services of a *demiurgos,* and have him create a vast universe, isolate from it his laboratory for immediate use, and discard all the rest of the universe as a left-over. Rather, he "designs" his spatial setting more or less to the extent, or to the degree of sophistication, to which he will employ it. And I think that it is this that Aristotle envisioned, farsightedly, though abortively.

What was limiting his vision was the sad fact that, when creating his notion of *topos,* he could only ask himself what is happening to the places occupied by wine and air when one pours wine from a jug into a goblet, instead of being able to ask himself what it means to say that the electric charge on a condenser accumulates spatially on or near the surface; or instead of being able to ask the much more intriguing question whether the "creation" and "annihilation" of elementary particles in contemporary physics is also accompanied by a creation or annihilation of individual *topoi* that are to be associated with them. Yet it is astonishing to see how much Aristotle was able to construct out of how little.

We stated above that his admirable aim was marred by a serious miss. Now this "miss" was his inability, in Book IV of the *Physica,* to make there a *total* and *complete* separation between (i) the definition of "space of the laboratory" and (ii) the notion of a cosmological space, which, by its very nature, must also be "gravitational" space, in the sense that it must also be a *natural* place for certain bodies to move toward, from some predetermined compulsion. This confusing feature was clearly recognized by Solmsen in his recent book.[7] In *Physica* IV, Aristotle's

[7] F. Solmsen, *Aristotle's System of the Physical World* (Cornell University Press, 1960), p. 128, also p. 443.

main characterization of the *topos* of a body (or of a physical system) is this, that it is "the inner boundary of what contains." This has nothing to do with gravitational properties at all, and Solmsen rightly observes that Aristotle was in an inarticulate fashion aware of this and actually hesitated to bring in the gravitational properties in the context of *Physica* IV; but he somehow brought them in nevertheless, though reluctantly. Solmsen also observes that the above characterization of *topos* is of no use to Aristotle anywhere else. We should point out perhaps that Aristotle does occasionally mention this characterization again, as for instance in *De Caelo* 326 b 8, but he does indeed make no use of it there in actual fact.

Another confusing feature is a seemingly serious inconsistency between the characterization of *topos* given above and an alternate one which follows in Aristotle not many lines afterward. We will deal with this inconsistency on its merits in a later section, but for the present we only wish to state that in contemporary physics the notion of space is beset with inconsistencies, and that no serious harm arises. Thus, (i) in the *ordinary* physics of our planetary system, space continues to be Newtonian; (ii) in the theory of single electrons or similar particles (that is, in the so-called quantum field theory), it is the space of special relativity; (iii) in the physics of our galaxy and beyond it becomes the space of general relativity; and (iv) the "statistical" space of quantum mechanics may be viewed as being different from, and thus inconsistent with, any "non-statistical" space, Newtonian or relativistic.

Finally we are bound to take note of the fact that Aristotle also mentions *topos* and *chronos* in his *Categories*, in Chapter 6 which is devoted to the category of "quantities." Whatever he says there about space is in the nature of logico-verbal statements and one might perhaps view it as a "dictionary definition." It pursues no particular aim, and one must not view it as intending to be a full-fledged alternate definition of space in rivalry with other definitions.[8] It would also be quite dangerous to attempt to date the *Categories* and *Physica* relative to each other on the strength of this discrepancy in the description of *topos* alone. Aristotle might have offered both descriptions "during the same semester" to different audiences, or in different contexts.

7. SPACE AND MATTER

In 209 b 11–13 Aristotle states flatly that Plato identifies his (cosmological) *chora* with *hyle* (matter), but in point of literal fact, Plato never does so. Many commentators, however, especially those in the 19th century and notably E. Zeller, agree with Aristotle that in a certain qualified sense in the *Timaeus* matter is space and space is matter. But others disagree. Taylor [9] for instance says to *Timaeus* 52 b 4—and Ross [10] is in complete accord with him—that matter really plays no part in Plato's cosmology at all, and that the "permanent implied in change" which to Aristotle is *hyle* is not thought of by Plato as "stuff" or "substrate" or "substrate of change." And Taylor concludes by saying that Aristotle "was probably unconscious that he was falsifying the theory of the *Timaeus* by forcing his own technical terminology into it."

To this we wish to recall from Chapter 2, that mathematicians, physicists, and scientists in general, when functioning qua scientists (and not qua "philosophers" perhaps) are incessantly perpetrating such "falsifications" of the most atrocious kind. Reformulations of someone else's results and interpretations in one's own technical terminology, even without much warning or any warning at all,

[8] For such views, see P. Duhem, *Le Système du Monde* (Paris, 1913), I, 197–200; also Max Jammer, *Concepts of Space* (Harvard University Press, 1954), p. 15.

[9] A. E. Taylor, *A Commentary on Plato's Timaeus* (London, 1928), p. 347.

[10] W. D. Ross, *Aristotle's Physics* (Oxford University Press, 1936), p. 566.

are almost taken for granted, and at any rate would not be branded as "falsifications," even when the emendations or corrections were intentional.

But even apart from the general attitude of laxity towards "falsifications" of this kind, there would be, to a physicist, a question in the present case whether there is any "actual" falsification at all. It is a fact that in present-day physics the word *matter* is used to designate not only the familiar "old-fashioned" matter of a very "tangible" kind but also over-sophisticated assemblages of so-called elementary particles which are the very opposite to being "tangible" in this sense. Now, these latter particles are only purely mathematical constructions in a sense. They are even more so than are Plato's "elements" in *Timaeus,* 53 C–55 C, and all properties of such elementary particles are totally inseparable from rather complex mathematical theories which are their setting and without which they would be beyond any possibility of description. Yet the physicist calls them "matter," not only from habit or indifference to appelation, but because, in their own theory, such "intangible" particles are indeed the ultimate elements out of which "ordinary" matter is composed. And since Plato fashions his elements out of figures which are structurally part of his *chora,* a young physicist of today would probably see nothing wrong or noteworthy in the finding that *chora* and *hyle* are "the same."

I would also like to make an observation which is only a side remark to the present immediate context. It is my view, which others may contest, that Plato's mathematical imagery in *Timaeus* 53 C–55 C is naïve and, relative to us, most "antiquarian," and that Aristotle's criticism of it in *De Caelo* III, far from being the "curse to science" that A. E. Taylor denounces it to be,[11] is to a large extent well taken.

It is true that Plato's general views on the nature of

[11] See n. 9, p. 403.

[157]

mathematics were on the whole perhaps more fit to subsume modern mathematics than were Aristotle's views. But these were only conscious over-all views of a philosophical tenor, and the differences between the views of Plato and Aristotle did not affect the actual "operational" fitness for bringing about a mathematical physics, say.

8. GRAVITATIONAL SPACE

It is a fact that Aristotle did not arrive at the law of inertia of Galileo and Descartes or at any laws of bodies falling under gravitation, even though physics of locomotion, both in the sublunary part of the universe and in astral reaches, was very much in his thinking, and a general notion of motion which was much more comprehensive than loco-motion proper was a central datum of his "chemistry," biology, and also psychology. Galileo continually casti-gated Aristotle for his inadequacies in (Galilean) me-chanics and, following his lead, commentators and in-terpreters are wont to take a most serious view of these defects and put great emphasis on them. But I think that it distorts history to burden Aristotle so pitilessly with a responsibility for some wrong guesses in physics which even the very great Archimedes did not right, and to allow this one aspect of failure of his to weigh so heavily against whatever else he may have achieved positively.

It is a most remarkable and extraordinary fact that the development of science after 1600 began with the estab-lishment of the so-called rational mechanics which held the stage of science in the 17th and 18th centuries, and in the 19th became the paradigm for all of physics and much of other sciences, which began to properly organize them-selves only then. And it is an equally remarkable fact that Greek science simply did not show a tendency toward a pattern of developments of this kind, but toward a pattern of developments of another kind in which no theory of mechanics was in the vanguard of advance. Aristotle, for instance, was a naturalist, and as such he had the relative standing of a naturalist of the 19th century, say. And in the 19th century, naturalists and also chemists were greatly attracted to thermodynamics, more than to any other field of physics or mechanics proper. Also, in the 19th century, a then prevalent version of thermodynamics, the so-called classical thermodynamics, was virtually independent of gravitational mechanics altogether. All this reacted on biology, and, for instance, the great treatise on 19th-century biology by D'Arcy Wentworth Thompson, *Growth and Form* (1917 and 1942), mentions the law of inertia but sparingly. It is a fact which, as such, does not speak against Aristotle that his thought patterns in physics are more reminiscent of syllogisms in classical thermodynamics than those of mechanics. Especially his notion of equilibrium in nature fits much more a thermodynamics of irreversible events than a mechanics of reversible processes. Most of the criticism of Aristotle by Galileo, Descartes, and other "mechanists" of the 17th century would not stand up before the forum of classical thermodynamics, if it rendered judgment solely within its own jurisdiction.

Finally, we wish to say that Aristotle's explanation of motion by *antiperistasis* (215 a 15; 267 a 16, 18), for which even fierce partisans of Aristotle feel apologetic, is in classical thermodynamics quite meaningful in a sense. In fact, in the latter theory "a sliding wall or piston will move only if, and as long as, the pressures on the two sides of the wall or the piston *in its immediate vicinity* do not cancel out." [12]

We also wish to state that even "atomists" like Epicurus or even Democritus did not actually show any tendency to

[12] I expressed this view in: "Aristotle's Notion of Place (Topos) in Physics," *Acts of the Xth International Congress of the History of Science, Ithaca 1962,* pp. 471–479.

initiate a mechanics of the 17th- or 18-century variety, say, although, on the authority of Archimedes, Democritus had quite a flair for mathematics too.

9. THE VOID (*Kenon*)

Aristotle himself motivates his denial of the existence of a void largely from an opposition to Atomism, but it is also a fact that in classical thermodynamics a "void" is envisaged but rarely, if ever at all. In such a thermodynamics, a physical system is made up of gases and liquids which are presumed to be contained in an over-all enclosure, to which there is no "outside," as far as the system is concerned. The notion of such an over-all enclosure does not allow for the presence within it of what one can call a "void." All that one could encounter would be a "vacuum" in some part of the system. And such a vacuum has to be created forcibly and artificially and is never "absolute." In the 19th century, all this was not deemed to be in disagreement witn the then basic assumption that all matter in chemistry and thermodynamics had a molecular structure. In spite of the molecular structure, for the derivation and application of large parts of the theory, gases and liquids were somehow assumed to be distributed continuously, with no "interstices" present.

10. PROCESSES-IN-NATURE

In present-day physics there are certain over-all notions like "physical system," "physical state," "physical process," "physical model," etc., which severally and jointly are an important background for the mathematization of physics. The Greeks never arrived at a stage of having notions that would adequately correspond to these. The nearest to this nexus of notions in the thinking of Aristotle was a certain notion of process-in-nature, but, as the notion appeared, it was too loosely broad and equivocal to become itself a background for a mathematization of physics, even if a tendency for such a mathematization had been present. In a rather forced interpretation one can say that the nearest to an effort for a would-be mathematization of the notion of a process-in-nature was made in Book V of the *Physica,* especially in Chapter 2, but it should be immediately said that this effort was a very abortive one.

Aristotle has various names for a process-in-nature, all of which already occur in Plato and earlier. The leading names are: *kinesis* (motion), *metabole* (change), *genesis;* and others are "alteration," "locomotion," etc. Commentators have been trying to put these various names into a hierarchical ordering, and various arrangements have been suggested, some of which embody alternate strands of ordering in one.[13] But none of the proposed arrangements is really convincing, and I venture to say that none is really possible. The fact is that the above conglomeration of names for the general notion of *kinesis* is fluid by origin, and such fluid conglomerations of notions in physics do not ordinarily solidify into sharply and meaningfully separated concepts, unless perhaps specific and definite meanings have been assigned to them by deliberate conventions. Another conglomeration of notions, which also causes great difficulties of interpretation, are words which are introduced in *Physica* V, 3, in order to express properties of linear ordering of a one-dimensional continuum (together, apart, touching, intermediate, successive, contiguous, continuous).[14] In this case the basic difficulty is this, that the Greeks at that time were not ready, or not yet ready, to think through with great precision something

[13] Compare J. P. Anton, *Aristotle's Theory of Contrariety* (New York, 1957), pp. 206–207; H. H. Joachim, Aristotle, *The Nichomachean Ethics* (Oxford University Press, 1950), pp. 269–275.

[14] Ross, *Aristotle's Physics* (see n. 10), p. 626.

which for modern mathematics, and in its setting, only the 19th century was able finally to do, although in this case the Greek achievements which are embodied in the fifth book of Euclid's *Elements* do have a virtually timeless validity.

With regard to the notion of *genesis* I wish to say that one must not over-dramatize its course of development. Classicists are prone to assert that *genesis* flourished under earlier Presocratics, was suddenly put under death sentence by Parmenides, was however resuscitated by Plato, and led a normal life under Aristotle. In physics, things somehow do not happen in this manner.

11. THE SCOPE OF PROCESSES-IN-NATURE

The notions of *kinesis* and *genesis* are not yet in Plato what they will be later in Aristotle. In Plato they are sometimes looser and broader, more indefinite and undifferentiated; or at other times, they are representative of a partial aspect or of a smaller phase of what they will be in Aristotle; or they are still enmeshed in Theaetetean logicality or Parmenidean existentiality, and are not yet fully textured in physicality as later in Aristotle. In Aristotle the allusions inherent in the notions are technically more "scientific" in our sense. If Greek "theoretical" science had continued directly into our science without first aborting in its own phase, the continuation would have become Aristotelian and not Platonic. It is sometimes suggested that Aristotle's enlargements of the Platonic notions of *kinesis* and *genesis,* say, were too small to be of great consequence, or even that they had a trivializing effect due to Aristotle's tendency toward putting notions into tighter frameworks, such tighter frameworks around Platonic notions having the effect of narrowing the width of their far-sweeping visions, as Platonists sometimes fear. It should be observed, however, that such tightenings of frameworks around notions in physics and mathematics are frequently of great importance, and seemingly small changes of nuance can be of considerable consequence.

In large parts of the *Physica* and in other Aristotelian contexts, *kinesis,* or one of the parallel notions, if representative of a general event, may represent what would nowadays be a physical, cosmological, chemical, or biological process; it may also represent one of many human activities, professional ones or, generally, intentional ones, as for instance the planning of the building of a house and the actual realization of the plans, even beginning with the inception of the intent and going through the necessary and natural stages of it up to and including the very completion of the task; or it may be the total process of painting a picture or carving a statue; or the process of acquiring a specific skill or a specific area of knowledge, or of educating oneself in general; or of carrying out a specific program of physical exercising for the sake of maintaining or improving one's health; or it may be the process of healing somebody or even oneself; etc.

However, one must not assume, after going through this catalogue of possibilities, that every conceivable "event" is a *kinesis* of some kind. In fact there are some peculiar limitations, on which Aristotle insists and about which he is fairly consistent. They are stated frequently in passages of his psychological works (*De Anima, De Sensu,* etc.), but the leading exposition of them is incorporated in his theory of pleasure in the *Nicomachean Ethics,* X, 1–5. An important and indispensable commentary on Aristotle's injection of the theme of *kinesis* into this context is in Joachim,[15] and further studies on this topic would be welcome. In the *Ethics,* what Aristotle wishes to exclude from being a *"kinesis"* is "pleasure." It apparently is not a "process," that is, something that leads up to and terminates in a "completion," but is a "completion" itself,

[15] See n. 13.

directly and immediately, not from some particular contraction or atrophy of what precedes a completion but
because there just is nothing at all that precedes. In a short
sentence, one can say that to Aristotle pleasure is not a
kinesis but an *energeia;* but there is something unsatisfac-
tory about making the sentence quite so short. The
important thing is that to Aristotle the nature of pleasure is
parallel to and even linked with the nature of sense
perceptions, like seeing, hearing, etc., which are also, as he
frequently and consistently maintains, not processes proper
but only completions. In support of this parallelism
Aristotle also adduces the fact that neither pleasure nor
sense perceptions require a specific interval of time for
their occurring, as any regular *kinesis* does. To this
argumentation it ought to be observed that in the *Physica*
the dissertation on time is given in Book IV and that in
various analyses of processes-in-nature in the preceding
three books time it is never mentioned and hardly alluded
to, although in later books of the *Physica* and in follow-
up treatises time does occur.

For a moment one might be tempted to say, when guided
by our own standards of "morality," that Aristotle wished
to keep "kinetistic determinism" out of the sphere of
ethical responsibility, but it probably would not be easy to
bring this into accord with Aristotle's attitudes towards
ethics in general or with pre-Hellenistic attitudes in
general. It is probably less of an improper modernism to
say that in linking pleasure to sense perception Aristotle is
aware of a certain "neurological" sameness in both, with
which one would certainly agree today. What would
probably be harder to agree to is Aristotle's view that the
"neurological" events which occur in both are not *kineseis*
proper but something else, that is, straightforward "completions." But one might agree with Aristotle to the extent
that the human consequences and concomitants of neuro-

logical processes are something other than physical processes themselves, and not entirely reducible to such.

12. OPPOSITES

An insistent feature of Aristotle's physics, which it is hardest to assimilate to present-day physics, at first sight at any rate, is the ever recurrent emphasis on, and reiteration of, the general theme of opposites, contraries, and contrarieties. Aristotle asserts and implies that the theme, as an active theme, was common to Greek rational physics from its earliest. In this he is ostensibly right, but he is probably overstating the role of opposites in earlier stages of development, as H. Cherniss has argued convincingly.[16]

Some of the opposites like "hot and cold," "dry and moist," "rarified and condensed" are to us common and ordinary, even too ordinary and commonplace and pre-scientific to be the central theme of a scientific development. Others, like act and potency, which are apparently due to Aristotle, do indeed have an originality of their own, but one wonders what one might be able to actively achieve with them in the end. Finally, a pair of opposites like privation and completion seem to be quite remote from us, and are even repugnant to our thinking, although, for instance, in biology the growth of an ovum after fertilization may be viewed as a "completion" resulting from the removal of a "privation."

It is true that opposites and contraries are significant and even recondite attributes of the universe around us. But Aristotle, and perhaps the Greeks altogether, stuck to the theme of opposites too doggedly and too unprofitably, or so it seems. In a retrospective evaluation, if such an evaluation be deemed admissible, one might be tempted to say that the over-intense preoccupation of Greek physics

[16] H. Cherniss, *Aristotle's Criticism of Presocratic Philosohphy* (Baltimore, 1935).

with the theme of opposites was the greatest single impediment to a mathematization of their physics, that is, to a development and unfolding of their science as it "should" have come about instead of as it actually did come about—abortively, that is.

13. MOTION OF MOTION (*kineseos kinesis*)

Nothing is seemingly more innocuous and ordinary than the opposites "hot and cold," "dry and moist," "rarified and condensed." And yet in the 17th and 18th centuries, while mathematicians and physicists were erecting their monumental edifice of "Rational Mechanics" with briskness and boldness, they were proceeding very slowly and with great circumspection in the area of thermodynamics, which is the area of physics to which these three pairs of opposites belong. The first truly "theoretical" results pertaining to these opposites were beginning to be envisioned only in the last decades of the 18th century, and only in the course of the 19th century were they actually codified.

Hot and cold are of course not genuine contrarieties, inasmuch as they can only be manifested by different positions on a thermometric or, at least, thermoscopic scale, with various intermediary positions between the two. As viewed in retrospect, there is in Aristotle a noteworthy attempt, an abortive one, at arriving at a would-be mathematical clarification of the role and meaning of positions intermediary between opposite or contrary ones. The attempt may be read into the argumentation in Chapter 2 of Book V of the *Physica.* Somehow, to me, Aristotle's doings in this chapter are crying out for a mathematical representation of a *kinesis,* as there investigated, by a function $y = f(x)$, $a \leqq x \leqq b$, where the endpoints $x = a$ and $x = b$ correspond to the starting stage and completed stage respectively, and the values x between a and b to the stages of the *kinesis* in between. If one views his *ki-*

neseis in Chapter 2 of Book V in this way, then all of Aristotle's statements about them are wrong, totally and disastrously so. In fact all his statements are negative statements, that is, warnings against what cannot or must not be done with the *kineseis* under consideration. As strictures on functions $f(x)$ they would prohibit the very operations with them without which there would not be any physics at all today, not even on the most elementary level. Aristotle's main contention (225 b 16–226 a 23) is this, that there cannot be a motion of motion, and he adduces for it no less than five proofs.[17] If applied to functions this would mean either that (i) there cannot be a function of a function; or that (ii) there cannot be functions of *two* independent variables; or that (iii) an "operator" like differentiation cannot be applied twice, and hence arbitrarily often, in succession. And the last alternative would mean that while one can form the notion of a "velocity" of a motion, which requires the formation of a first derivative, one cannot iterate this to form the derivative of the velocity, which is "acceleration."

Now, it is precisely such an incapacity to conceptualize the rate of change of the rate of change which separates Archimedes from Newton by a threshold which Archimedes never succeeded in crossing.[18] Purely mathematically, there is nothing in Newton's *Principia* that was not familiar to Archimedes, except the notion of the rate of change of a velocity. And even here, only the notion was alien to Archimedes, and not the power for formalizing the notion mathematically, if by some reversal of history it had come within his purview. In fact, purely mathematically Archimedes was much better equipped for dealing with it

[17] See the translation by H. Carteron in the Collection Budé, *Physica*, II, 15–17.

[18] J. P. Anton (see n. 13), Appendix 1 to Chapter 4, already stated that Aristotle's denial of change was in a sense reversed in the mathematical calculus of the seventeenth century, especially in Newton's *Principia*.

formally than was Newton, seeing that Newton did not manage to learn to define really rigorously the notions of velocity and acceleration to the very end of his days.

Aristotle's statement (225 b 15–16): οὐκ ἔστι κινήσεως κίνησις οὐδὲ γενήσεως γένησις, οὐδ' ὅλως μεταβολῆς μεταβολή (there cannot be motion of motion or becoming of becoming or in general change of change) is a devastating self-indictment of Greek rational thinking at its root. It decrees that, in general, Greek thinking would not, because it could not, form a relation of relation, or a property of properties, or an aggregate of aggregates. This is precisely the kind of limitation of Greek rationality which we have described in the last section of Chapter 1, and we will now adduce further manifestations of it.

Archimedes did not have coordinate systems or mathematical functions. Nevertheless, in his book *On Spirals* he came quite close, in his own setting, to forming the derivative of a function

$$\frac{dy}{dx} = \frac{df}{dx} \tag{1}$$

which is the mathematical prerequirement for the "abstract" conceptualization of the notion of velocity. However, in order to advance to the concept of acceleration, one has to be able to form a second derivative. This requires that one form the derivative (1) at each and every point x, then view the resulting mathematical object as a new function in x, and then apply the "abstract" process of differentiation to this new function again. It is this kind of iteration of logico-ontological abstractions to which Greek thinking was never able to penetrate to any noticeable extent.

Or, take the general (mathematical, or logical) notion of a "set" or "aggregate" as Georg Cantor introduced it (a "set" being a multitude of "elements" about which nothing is assumed known except that any two elements can be distinguished from each other by the separateness of their identities). Such a notion of "set" does not occur anywhere in Aristotle, and it was not congenial to his thinking, but it may nevertheless be conceded that this notion would have been within the easy grasp of the Aristotle of the *Prior Analytics,* say. However, in order to "operate" with his sets, Cantor *very quickly* forms, or constructs, or envisages, sets of sets, that is, sets whose elements are themselves sets; and most of his operations with sets apply to such. Now, it is this kind of "iteration" of the notion of sets of which only the barest rudiments occur in the *Prior Analytics,* or anywhere else in Aristotle or his successors, if they occur at all.

The most striking application of sets of sets is involved in Cantor's proposition that with any cardinal number whatsoever one can associate a second cardinal number which is genuinely larger than the first. Cantor takes an arbitrary set A, and he conceives of the set B whose elements are the subsets of A; and, by a reasoning whose originality will never dim, Cantor then argues that the cardinal number of B is genuinely larger than the cardinal number of A. It is this kind of concept formation and reasoning which lay outside the known ontology of the Greeks. In this sense, the universes of Parmenides, Plato, Aristotle, and Archimedes were tightly closed, with "no exit."

Plato's procedure of logical differentiation which goes under the name of "dichotomy," although it embodied a sequel of steps, did not embody a sequel of abstractions. It was vertical; but it moved vertically downward and not upward, and it was a succession of logical particularizations and not of abstractive self-generalizations. Plato did have an abstractive process which moved upward, namely the process of abstracting a (Platonic) Form or Idea from its underlying realizations. However, a Platonic Idea, when viewed as an abstraction, was a terminal one, and no further abstraction could be superimposed on it; or, to put it differently, once the Platonic Ideas were arrived at, there was nothing that Plato, or anyone else, could actually "do" with them. In *Metaphysica,* Book XIII, one of Aristotle's arguments against Ideas was the assertion that Plato had not provided how, out of the *Idea* of three-ness and the *Idea* of four-ness, to form (by an "abstract" mathematical operation with Ideas) the *Idea* of seven-ness. In this argumentation Aristotle was irrefutably right; but we have to add that Aristotle's own advice of not forming Ideas at all was a council of despair, and made matters very much worse.

We also wish to point out that the Greeks were never driven to conceptualizing, or rather to formalizing, the notion of "relation." In the 20th century, this notion is a most primary datum, and no fruitful thought can be conceived without it. The Greek antecedent of our "relation" is their *pros ti*. The "Dictionary" in the *Metaphysica* has a long entry about it (1020 b 25–21 b 12), but the entry is not inspiring reading. The Greeks did not make their *pros ti* into a substantive "category" for "operational" rationalization, as it is with us indispensably today; they even did not make it into some serious linkage within "logification" of knowledge, as some medieval Schoolmen may have attempted to do, although not successfully.

14. SIMPLE AND COMPOSITE CHANGES

As we have already stated above, the *kineseis* of Book V of the *Physica* are presumed to be such as traverse various states which are intermediate between terminal ones. In contrast to this, Book I envisages various changes (*geneseis*) in which states intermediate between the terminal ones do not occur at all. Such changes are indeed very "simple," and in the Loeb edition the "argument" preceding Chapter 7 interprets the text to imply that "change, however complex, can always be resolved into a combination of antithetical movements." [19] The text of the *Physica* does not quite say so; but the interpretation is a plausible one, and we will adopt it for the moment for the sake of an observation to be based on it. In present-day physics the decomposition of a physical process into "simple" processes each of which is an indecomposable one-step event (any so-called emission or absorption, or alternately, creation or annihilation of elementary particles) is in the very center of the entire physical interpretative thinking.

Altogether I would say, though with some reservations, that two-valued truth functions of logic and the succession of yes-or-no signals which constitute a context of information in a contemporary computer would have been more in accord with Aristotle's outlook on physics in Book I of the *Physica* than would be the mathematical apparatus of point functions in Euler–Lagrange mechanics, say, although the latter would have fitted into the outlook of Book V, in a sense.

In sum I would say that 20th-century developments of science in general, and of physics in particular, are more compatible with Greek preferences for basic two-step physical processes and for basic pairs of opposites and contraries than were developments in the 19th century. Also, for instance, a resolution of every neurological process into a succession of "choices" between inhibition and stimulus, which would have been just as congenial to Aristotle as it is to a 20th-century scientist, might have sounded quite alien to somebody in the 19th century. However, these 20th-century developments are not self-sufficient, but are an outgrowth of past and present developments of "continuous" processes and are surrounded by them, whereas in Aristotle the two-step processes were meant to be grounded in themselves. Thus, in retrospect one can say that in this as in

---

[19] I, 69.

other similar cases the Greeks were rushing ahead too fast too early with too few resources at their command.

15. ARISTOTLE'S THEORY OF *Topos*

We have already stated in section 6 that Aristotle's *topos* in *Physica* IV is meant to be a "spatial" setting for his pre-notion, such as he had, of an individual physical system, and that, for an understanding of his purposes and achievements, his half-hearted attempts, against his better judgment, at also injecting "gravitational" properties of natural place into the theory of Book IV ought to be ignored.

After a lengthy deduction from would-be axiomatic demands, Aristotle arrives at the characterization (212 a 5–6), already stated above, that if a physical system is enclosed by another, the *topos* of the system is "the innermost boundary of the enclosure." The context of the *Physica* also shows [20] that this concept of *topos* also implies a certain aspect of "voluminousness," but not a point-by-point "extension" which links boundary elements across the bulk of the system. This characterization of space obviously does not fit a mechanical system of finitely many mass points, because in such a system only point-by-point localization can be realized. However, this characterization begins to fit a system of hydrodynamics or electrodynamics, say, because frequently, for such a system, its equilibrium or alterations of equilibrium are determinable from events located on the "innermost" part of the boundary layer. And this characterization fits best a system of gaseous substances in classical thermodynamics, say, because the so-called characteristic equations of the theory involve spatiality only through the volume but through no other structural determination of space; in particular this characterization makes no reference to

[20] Taylor (see n. 9), p. 676.

points or figures in space, nor even to the three-dimensionality of it. We also note that the organization of a living cell in biology is largely determined by events across innermost parts of "boundaries" which in cell biology are called "membranes." [21]

After stating the above characterization of *topos,* Aristotle continues to analyze the phenomena which led to it, and he seemingly involves himself in inconsistencies. He argues as follows. If a boat moves on a river which is itself in motion, then, by the above characterization, the *topos* of the boat ought to be the innermost boundary of the water which surrounds the boat. But since this water is changeable, Aristotle somehow does not like to permit it to define a *topos* of the boat; and Aristotle suggests that such a boat does not really have a *topos* in its own right, and that only the river as a whole has a *topos.* Altogether he then proposes the alternate characterization (212 a 20–21) that "place is the innermost *motionless* boundary of what contains."

Some commentators, ancient and modern, find the two alternate characterizations of *topos* irreconcilably contradictory.[22] Some recent ones, however, find that the second characterization only superimposes a "Principle of Relativity" on the first, and that in the end there is no inconsistency at all.[23] In my own view the inconsistency can

[21] See, for instance, L. L. Langley, *Cell Functions* (New York, 1961), Chap. 3 and other passages.

[22] See Duhem (n. 8). We note that Duhem was altogether very critical of any kind of inconsistency in physics. Thus his book: *Les Théories électriques de J. C. Maxwell; Étude historique et critique* (Paris, 1902), was written mainly with a view to exposing various inconsistencies in the sum total of the electromagnetic work of Maxwell.

[23] The most detailed interpretation of this kind was given by B. G. Kusnezow, "Die Lehre des Aristoteles von der relativen und absoluten Bewegung im Lichte der modernen Physik," published in the book *Sovietische Beiträge zur Geschichte der Naturwissenschaften,* Herausgegeben by Gerhard Harig (Berlin: Deutscher Verlag der Wissenschaften, 1960), pp. 27–61.

be better resolved, although not absolutely so, by another interpretation which I think is more "informal" and more suitable to Aristotle's approach than the Principle of Relativity is. Instead of exemplifying our explanation on Aristotle's boat we will take another case which is equally "ordinary."

Suppose that in a workshop we observe an iron pipe as to its common properties. To begin with, its "place" would be the ordinary "extension" it displaces. If, however, the pipe happens to be rusted and we want to account for this too, still in an everyday manner, then we have to consider that rusting is combination with oxygen; and assuming that the oxygen in the workshop circulates freely, then the oxygen of the entire workshop is responsible for it. Thus, even if the state of the iron pipe alone is at issue, for the examination of the rust the place of the object must be enlarged to extend over the whole shop, in a sense. Furthermore, if, say, some magnetization of the pipe has been observed and it cannot be explained by events and installations in the shop itself, then, for the sake of an explanation, the "true" spatial setting of our object must be even further enlarged.

Returning to Aristotle's boat, we might elaborate on his own argumentation as follows. If we stay in that boat below deck and are only concerned with events inside it, then the genuine place of the boat is determined by the interior boundary of the water flowing past it. If we become concerned with the effect of the water on the condition of the hull, then the entire river becomes the spatial setting, and if we wish to consider even the effect of precipitations and winds on the level and behavior of the water, which in its turn affects the hull of the boat, then the atmosphere above the river must also be added.

Actually, Aristotle himself says (209 a 32–35, Oxford translation): ". . . you are now in the heavens because you are in the air and it is in the heavens; and you are in

[174]

the air because you are on earth and similarly on the earth because you are in this place which contains no more than you."

The above interpretation is the elaboration of a previous one.[24] In Chapter 6 we will give an alternate elaboration, which is of a more mathematical nature and somewhat different from the present one, but compatible with it.

16. THE UNIVERSE AS ONE PHYSICAL SYSTEM

Of the five chapters on the theme of *topos* in Book IV of the *Physica* the last one, Chapter 5, is inconclusive but intriguing. As we have repeatedly stated, we are adopting the interpretation that *topos* in the *Physica* is meant to be (non-gravitational) place pertaining to an individual physical system. In the *De Caelo,* however, *topos* is, quite differently, the gravitational space of the *kosmos* as introduced in the *Timaeus,* more or less. Now in Chapter 5 Aristotle seems to be raising the following question. What happens if one views the whole universe, τὸ πᾶν, as one vast though "individual" physical system, that is, if one views the "whole" not as the *kosmos* of the *Timaeus* but as an individual physical system which happens to be coextensive with the entire universe? This, incidentally, is a very legitimate "scientific" question. Aristotle puts it somewhat vaguely, but discernibly, and it does not belong to Parmenidean dialectic, let alone to some kind of "sophistry," but to physics proper. Aristotle asks whether this comprehensive physical system also has a *topos* of its own, or not.

Aristotle has no solution to this problem, which is indeed a vexing one, but he ponders it and throws out fragments of thought about it. He thinks that the situation is ambivalent, and that in some respects *to pan* does have a *topos* and in others not. Perhaps one ought to distinguish

[24] See n.12.

between being in place "actually" or only "potentially," he suggests. This is a routine suggestion of his, and he observes apropos of this that one might say that in a homoeomeric substance the constituent elements have mutual places potentially, whereas in a granular substance they have them actually. Or, he thinks, even if the whole is viewed as one physical system, it might still be pertinent to subdivide it appropriately into parts and then assign a *topos* to each part, although this cannot be done for the whole itself. Or, one might distinguish between something having *topos per se* or only *per accidens* and then decide that the whole has it only *per accidens*.

After offering such suggestions for a stretch (212 a 31–212 b 23), he suddenly runs out of possibilities, and in a closing statement (212 b 23–30), which is also a summary, he avers that he has achieved all that he had set out to do.

In Book I of *De Caelo,* which continues the *Physica,* Aristotle expounds his doctrine that the world (οὐρανος) is finite. He is having difficulties with distinguishing between finiteness of matter and finiteness of the geometric substratum, and with determining what finiteness of either actually means. However, this is the kind of difficulty with which contemporary cosmology, on its own level of sophistication, has also to contend, fully so. Also, a scientific cosmology does not necessarily convince by logical neatness. There is too much at stake to be persuaded by a smoothly flowing phraseology, verbal, logical, or even physical.

For reasons which present-day physics might not find it easy to endorse, during and after the Renaissance much criticism was heaped on Aristotle on account of the doctrine. But it is good to remember that Johannes Kepler, the pacemaker of mathematico-physical astronomy of today, stoutly championed the cause of Aristotle; that Isaac Newton never uttered an unkind word about Aristotle, whatever Galileo, Descartes, and even Leibniz [25] may have said; and that in the 20th century the cosmological work of Albert Einstein has made "Finiteness of the Universe" a very fashionable doctrine again.

17. ACT AND POTENCY

By a somewhat forced analogy even Aristotle's pronouncement in *Physica* III, 1 that *"change is actualization of the potential"* or, more literally, that *"the fulfillment of what is potentially, insofar as it exists potentially, is motion"* can be made to correspond vaguely to something important in modern physics, although in the 17th century Galileo, Descartes, and apparently also Leibniz [26] were bluntly asking of what possible use this pronouncement could be to anybody, ever.

To myself the pronouncement evokes the imagery of the fact that in a "natural" and "free" mechanical process potential energy is transformed into kinetic energy, and that the sum of the two is conserved. And if one permits this analogy, then one can also forgive Aristotle for having reiterated his pronouncement with a somewhat wearying insistence. In the 19th century the theorem on the transformation and conservation of energy was comprehended by scientists but very gradually, and only after the most wearying repetitions of its meaning and purpose by the sponsors of the theorem; whereas in the 20th century general theorems of this kind are taught to a student in physics briskly and in short order, and then quickly applied to situations of gravest import. Even the great Max Planck, before introducing his quantum hypothesis at the age of 42, spent the anticipatory and formative part of his scientific career in efforts to elucidate the thermodynamic datum of energy as a general conception of physics. This was highly

[25] A. Koyré, *From the Closed World to the Infinite Universe* (New York, Harper Torchbook, 1958), pp. 58ff.

[26] Léon Robin, *Aristote* (Paris, 1944), p. 300.

Aristotelian, in a sense; and it was this which gave Planck the successorship to Kirchhoff at the age of 31, which was an academic distinction of magnitude at the time.

It should also be stated that although Aristotle's denial of the possibility of an *actual* infinite is open to argumentation, his drawing up of a distinction, with emphasis, between potential and actual infinitudes is one of the most durable specific achievements of logico-mathematical thinking ever.

18. CONCLUSION

If one compares the physics of the Greeks with our theoretical physics of the last few centuries, and if one does so pragmatically and without condescending sentimentality, one may come to the conclusion that Greek physics hardly made a start. Except for the two laws of Archimedes, namely, the law of the lever and the law of the floating body, which Archimedes and others viewed as mathematics proper, Greek physics did not state a single law which could be put into a present-day textbook in the manner originally stated. Greek physics did in no way bequeath a book which would compare to works by Euclid, Archimedes, or Apollonius in mathematics. Yet Greek physics produced the system of Aristotle, which, although Aristotle was a naturalist rather than a physicist, held the stage of physics for almost 2,000 years, and which, by its flashes of insight and uncanny anticipations, evokes fascination even today. And those who scoff at Aristotle ought to tell us, or tell themselves, where we would be today if none of his works had survived.

CHAPTER 5

THE ROLE OF MATHEMATICS IN THE RISE OF MECHANICS*

MECHANICS, as a part of physics, is very mathematical, in imagery, content, and consequences. We will make some observations on the mathematical structure of mechanics during some phases of its development, and we will be concerned with the inward mathematical texture rather than with the outward mathematical setting. But we will not go essentially beyond the beginning of the 19th century. After that time, mechanics and paradigms from mechanics began to penetrate into many other areas of theoretical physics, so that observations on the mathematical nature of mechanics after the early 19th century would have to take large parts of theoretical physics into account. In the 17th and 18th centuries mathematics and mechanics were developing not only in constant interactions but also in close intimacy. But, in the 19th century, relations between mathematics and mechanics, or rather between mathematics and theoretical physics, began to be different. The interactions continued to be no less in importance, but they began to manifest themselves in parallelisms of development rather than in direct interpenetrations, and an analysis of their new relations requires categories of description other than those applicable to previous stages.

Modern mechanics came into being soon after the Renaissance, toward the end of the 16th century, and its stages of development, taken by centuries, have been as follows. The 17th century in mechanics extended from

* Originally in *American Scientist,* Vol. 50 (1962), pp. 294–311. Slightly revised.

Stevin to Newton and culminated in Newton's *Principia*. The 18th century in mechanics extended from Newton and Leibniz to Lagrange and culminated in his *Mécanique analytique*. The 19th century in mechanics extended from Lagrange and Laplace up to the first stirrings of quantum theory in 1900, and in a certain qualified sense it culminated first in Maxwell's *Electrodynamics,* inasmuch as the latter theory imitated hydrodynamic wave theory, and secondly also in leading treatises on statistical mechanics. In the 20th century the leading new theories of mechanics have become, more than before, direct and full representatives of the new physical theories of which they are a part and can hardly be isolated from them for separate analyses.

We venture to characterize the past centuries of modern mechanics by the following headings. The 17th century was an age of revelation; the 18th century was an age of patristic organization; and the 19th century was an age of canonical legislation. If we dared to continue we might suggest that the 20th century is an age of reformation, or perhaps an age of reformation and counter-reformation in one.

1. BEFORE THE 17TH CENTURY

In Greek antiquity, Hellenic and Hellenistic, if viewed in retrospect and if compared with the 17th century, say, there seemingly was an imbalance between mathematics and rational mechanics. The Greeks had a systematic mathematical astronomy, but outside of this they seemingly did not arrive at a theoretical mechanics or theoretical physics that was equal in status to their mathematics. Their mathematics, whatever its limitations, had the size and organization of a system. Their mechanics however was only an assortment of achievements, whatever their eminence.

Some of the difficulties that kept Greek mechanics from developing into a system may have been linked to peculiarities which kept Greek mathematics from advancing beyond the stage at which it was arrested. Anybody who appraises the mechanical work on the balancing of a lever in Archimedes and in his successors, Hellenistic and medieval, cannot fail to notice that in some covert and tacit manner each of them had the concept of the statical moment in his reasoning, but that none of them reached the point of formulating the concept expressly as the mathematical product $P \cdot L$ in which the factor L represents the length of an arm of the lever and the factor P represents the size of the weight which is suspended from it. Archimedes had a pre-notion of the moment in his context, but something in the metaphysical background of his thinking barred him from conceptualizing it overtly and "operationally." Many physicists after him—including Heron, Pappus (A.D. 320), Jordanus (13th century), and Leonardo da Vinci (15th century)—were groping for some articulation of the notion of the moment; but for over 1,900 years it kept eluding them, until finally Isaac Newton gave a clear-cut formulation of one. It was not the statical moment of rotation, but the inertial moment of translation, and definition 2 of Book I of the *Principia* introduces it thus: "The quantity of motion is the measure of the same, arising from the velocity and quantity of matter conjointly."

The express formation of moments became a great leaven in modern mechanics, and the introduction of moments into electrodynamics has been a most creative aspect in the unfolding of modern physics from its core. This makes it the more intriguing and the more urgent to inquire how it could come about that the statical moment of the Greeks—which they very nearly had—was not expressly conceptualized long before the 17th century and especially why Archimedes himself, great intellect that he was, was not forcibly led to introducing it. Now, a historical question like this may be deemed to be methodologically not admissible. But, in fact, if it be admitted, then a specific answer to it quickly suggests itself. The answer is this, that even if Archimedes the physicist had "wanted" to introduce the statical moment expressly, Archimedes the mathematician would not have been able to do so. If L_1 and L_2 are two quantities of a certain kind, say both lengths, and P_1 and P_2 are two quantities of another kind, say both weights, then Greek mathematics could envisage the proportion

$$P_1 : P_2 = L_2 : L_1$$

and decide whether it does or does not hold for special values of the quantities at hand. But it could not convert the proportion into an equality

$$P_1 \cdot L_1 = P_2 \cdot L_2$$

and verify its validity in this form. Greek mathematics could not or "would" not form a product $P \cdot L$. If P and L are general real numbers it could not form the product because it simply did not have real numbers. And if P and L were two general magnitudes, even if it did "have" the magnitudes separately, it still would not conceive of forming their product, especially if the factors P and L represented magnitudes of heterogeneous kind as weights and lengths are. And so Archimedes the physicist and generations of physicists after him could not have a statical moment in mechanics as they should have had.

In a certain sense, and again as viewed from the 17th century, the quick answer we have just given does no more than raise another question in the place of the first. If all that was wanting was the mathematical conceptualization of a product like $P \cdot L$, then why was Archimedes, the great mathematician, not driven to doing just that? To this question, if it be not ruled out as inadmissible, there is no quick answer. In fact, in the 17th and 18th centuries, if mechanics edged up to a conceptualization which the then mathematics could not articulate then mechanics would simply bid mathematics to learn to do so, and mathematics, after having learned to do so, would react on mechanics in return by suggesting to it more than originally required of it. In other words, mathematics and mechanics were then erecting and exploiting parallelisms and correspondences between mathematical and mechanical entities, conceptions, and articulations. The opening pages of Newton's *Principia* are impregnated with such parallelisms. They exasperate expounders of the *Principia* who insist on hierarchical orderliness and unequivocality of meaning in depth, but they also generate the vigor and the speed with which the stream of assertions and implications rushes by.

S. Sambursky, in his recent book on *Physics of the Stoics* (1959), rightly observes in passing that Stoic attempts at building up a mechanics of continuous media ultimately failed because, among other reasons, mathematics did not aid them. In this Professor Sambursky is right. But we wish to qualify his finding by adding that it was a fault of the Stoics themselves if mathematics did not come to their aid because, to a Stoic physicist, Archimedes was a mathematician, and a mathematician only, and not somebody who could be a mathematician and a "theoretical" physicist in one.

The Greeks did not themselves introduce mathematics as a technique for mastering problems that arise in the consciousness and thinking of man. This had been done by others before, as the Greeks themselves fully acknowledged. But they did recognize in mathematics a universal arche of knowing. They so proclaimed it to be and they would not have allowed this universality to be qualified by differences of eras and of philosophical settings. We think therefore that it is legitimate for us to probe, even in retrospect and even against the different philosophical spirit of the 17th century, the extent to which the Greeks themselves succeeded in realizing and exploiting the force and effectiveness of this universality.

Archimedes also accomplished basic work, perhaps his most famous one, in the mechanics of floating bodies. Here again he did not introduce the physical concept which is central to the subject matter, namely, the concept of hydrostatic pressure. But in this case Archimedes may be "excused." Modern mechanics had great difficulties in conceptualizing the notion of pressure, although Stevin immediately mapped out the task of doing so, and almost everybody after him pursued it. Even Newton was not yet quite certain of it. In a sense the clear-cut mathematization of the concept of pressure was arrived at only in the course of the 19th century, beginning perhaps with work by A. Cauchy on equations of motion for a continuous medium in general. In the 19th century the mathematical "image" of pressure became a tensor, albeit a very special one, and the actual formalization of the concept of tensor and a full realization of its mathematical status took a long time to emerge.[1]

---

[1] We recall footnote 46 in Chapter 2, in which we have stated that a general tensor in three-dimensional space is determined by a square array of 3-by-3 numbers,

$$a_{11}, a_{12}, a_{13}$$
$$a_{21}, a_{22}, a_{23}$$
$$a_{31}, a_{32}, a_{33}$$

If hydrostatic pressure is viewed as a tensor, it is a particular case of a so-called stress tensor; and a general stress tensor is a general tensor which is symmetric, and thus requires six numbers for its determination.

Archimedes dealt with what is called a *perfect* fluid, and in this case the stress tensor has the following special form. The components which are not on the diagonal are zero, that is,

$$a_{12} = a_{13} = a_{21} = a_{23} = a_{31} = a_{32} = 0$$

so that only the three components

$$a_{11}, a_{22}, a_{33}$$

The Schoolmen of the 13th and 14th centuries were inspired to attempt to conceptualize notions from mechanics like velocity, acceleration, and force, by introducing corresponding elements of the infinitesimal calculus and by even trying to formulate coordinate geometry. It has not yet been fully established to what extent these medieval efforts were a truly anticipatory phase of 17th-century developments. The fact is that in the 15th and 16th centuries the pace of mechanics slackened, whereas mathematics was quite active, but in directions of its own. By the end of the 16th century, mechanics somehow became very vital again. It showed great vigor, acted as if it had made a fresh start, and suddenly there existed a Rational Mechanics in the full sense of the word. Mathematics was quite active during the 16th century, but after 1600 it quickened its tempo even more and also entered on developments which were different from those of the preceding century.

During the Renaissance period, mathematics was preoccupied mainly with what is commonly called arithmetic and algebra. Now, whatever the origins of this mathematics may have been, about the effects of it there can be no doubt. It brought about momentous innovations which gradually wrought a changeover from the inertness of traditional syllogistic schemata of Greek mathematics to the mobility of symbols and functions and mathematical

---

remain; and, moreover, these three components have equal values. Thus, there is a value p ($=$ pressure) such that

$$a_{11} = a_{22} = a_{33} = p$$

Therefore, Archimedean pressure is determined by a single quantity, namely p. Nevertheless, in present-day mathematical hydrodynamics, pressure is not viewed as an "ordinary" scalar, but it is represented by a tensor which happens to have the very special form

$$p, \ 0, \ 0$$
$$0, \ p, \ 0$$
$$0, \ 0, \ p$$

relations. These symbols and functions and relations have penetrated into many areas of systematic thinking, much more than sometimes realized, and they have become the syllables of the language of science. What mathematical symbols and functions are nobody can tell, other than in tautologies, that is. The Greeks found that mathematical thinking has a universality *sui generis,* and investigations into foundations of mathematics are not much more than systematic explications of what this implies.

The Renaissance age also initiated the theory of probability. From being a gambler's calculus of finitely many combinatorial possibilities it eventually developed into an analytical theory of laws of large numbers and of infinitely many possibilities in discrete and continuous distributions. During the past 100 years, Probability entered Mechanics in a major way, and it did so twice. In the second half of the 19th century it created statistical mechanics for the kinetic theory of matter; and in the 20th century it was called upon to mediate between two conflicting approaches to quantum mechanics, corpuscular and undulatory. The mediator's judgment was unprecedented and at first quite disturbing.—Also, as if in return, a phenomenon from mechanics which was first observed by a naturalist in the early 19th century, the so-called Brownian motion, has in the 20th century stimulated the emergence of the probabilistic concept of a general stochastic process. The concept is of undetermined definitory scope, but it has been in the center of probabilistic thinking for several decades past.—Except perhaps for combinatorial speculations, there are no records of quantitative probabilistic preconcepts in Greek antiquity. There is a provocative section on Chance and Fortune (or rather Chance and Fortuitous) in Aristotle's *Physica* (Book II, Chapters 4, 5, 6), but regrettably it has no relevance to Mathematics or Physics as we understand them. There seem to be no antecedents to modern Probability in earlier eras and civilizations, and in view of Probability's link to Chance and Fortune this is most remarkable.

2. THE 17TH CENTURY

The 17th century produced the greatest single statement on the relation between mathematics and physics. It is Galileo's dictum that mathematics is the language of science. It is more than an aphorism. It asserts that although mathematics and physics are inextricably linked they nevertheless have separate identities and are hierarchically not comparable. Attempts to bridge the separateness of their identities and to put them into hierarchical dependence have been undertaken frequently.

Attempts to subordinate mathematics to physics (or even mechanics) usually center in the statement that "geometry is a branch of physics." On the other hand, certain contemporary trends, especially those which can be traced to Leibniz' plan for a universal *mathesis,* may be interpreted as giving to mathematical knowledge and to logico-mathematical formulations a supremacy over the realm of science, not only natural science but even social science including perhaps history too. Leibniz' plan had some antecedents in the 14th and 15th centuries. Also, some Schoolmen of the 13th century, above all Roger Bacon, have become known for their esteem for mathematics, but it would not be easy to attribute to them a reasoned metaphysical judgment on the relation between mathematics and physics.

In Greek antiquity, according to Aristotle, Pythagorean circles maintained that in nature "all is number," meaning apparently that "all is mathematics." In connection with this, some of them also maintained that space, meaning apparently the space of physical events, is made up of discrete points. Before deprecating the degree of sophistication of this alleged Pythagorean view, one should realize that over 100 years ago the mathematician, Bernhard

Riemann, in a famous address on differential geometry, also suggested that space might turn out to be discrete. Plato's doctrine or doctrines also incline in general to give to mathematics a supremacy over much of other knowledge. However, Aristotle also reports, with some emphasis, that Plato assigned to objects of mathematics a hierarchical position intermediate between sensibles and pure Forms (Ideas), which obviously circumscribes the supremacy of mathematics and, what is more important, gives it an identity of its own. This was a very fine glimpse of the true nature of mathematics, but Aristotle, within his general condemnation of Plato's theory of Ideas, battles this Platonic view and advances instead his own that mathematics is an abstraction from nature. Aristotle's view clearly does not fit modern mathematics. If π is, as usual, the area of the circle of radius 1, and e is the basis of natural logarithms then the number e^π is hardly an abstraction from anything in nature, although π is, and e by an expenditure of casuistry could be made to be so too. Also the complex number $i = \sqrt{-1}$, although a solution of the equation $x^2 + 1 = 0$, is certainly not an abstraction from anything in nature, and yet complex numbers were used successfully by A. J. Fresnel in the theory of refraction in 1818 and have spread into all parts of physics for many purposes.

In the 17th century almost everybody excelled in more than one area. Galileo the physicist had a standing as philosopher. Pascal was scientist and *penseur*. Leibniz was philosopher, mathematician, logician. Kepler was astronomer and mathematician. Descartes was evenly divided between philosophy and mathematics. And Newton was equally great in mathematics and in physics. A notable exception was Huygens. He was only a physicist. But he was a physicist's physicist, although he did not even write a treatise.

Some leading works of the 17th century give the

impression of continuing, by direct succession, works of a distant past, thus spanning gaps of long stretches of time. For instance, Galileo does not take much notice of 14th-century Schoolmen, but he acts as if he were a direct successor to Aristotle and he "corrects" Aristotle on specific statements made in the *Physica*. Descartes in his geometry acts as if he were a direct successor to Apollonius (210 B.C.) and Pappus (A.D. 320), and in a certain sense he even taunts these "predecessors" of his for not having achieved themselves what he is now about to do. And it is indeed inexplicable why Greek mathematics which lasted over 1,000 years, namely, from 600 B.C. to A.D. 400 and beyond, and which had a geometric tenor did not arrive at such simple things as placing figures into pre-introduced coordinate systems and plotting curves on graph paper. Fermat, who, by mathematical passion, was mainly a number theorist, considered himself to be direct successor and disciple to the ancient number theorist Diophantus (A.D. 250). He revered him, and as a gesture of reverence he even quoted Diophantus in a non-arithmetical context in a paper on maxima and minima (Fermat, *Oeuvres,* Vol. I, p. 133). But the most notable instances of such "leaping backward" are in the achievements of Kepler and Newton.

Kepler was an assistant to Tycho Brahe, a disciple to Copernicus, and a spiritual descendant of Nicolas of Cusa (15th century), who is commonly considered to have been the harbinger of an important phase of modernism. These are straight facts and could not be gainsaid. But if, from a distance, and in retrospect, one elects to look at the bare mathematics and physics only, and only at the sharp lines in the contours of development, then one may also find that Kepler's discipleship goes back directly to Apollonius and also Ptolemy (A.D. 150) and that Kepler was a successor directly to them. To generations of astronomers, from Eudoxus (370 B.C.) to Copernicus and Tycho Brahe, the basic celestial figure was a circle, and they took it apparently for granted that a celestial trajectory is a figure composed of circles kinematically. Kepler broke with this supposition. He introduced ellipses instead, and, guided by astronomy, he grasped the true meaning of foci of conics which others before him had no occasion to recognize. Conics, and their theory, were in no way Kepler's private mathematical invention and discovery. They had been in the public domain for nearly 2,000 years for anybody to find and use. Also, in point of literal fact, Kepler did not find his ellipses in the tables of Tycho or the "revolutionary" writings of Copernicus or the "learnedly ignorant" utterances of Cusanus. He found them by searching, untiringly, in the work of Apollonius, and by being, apparently, also thoroughly familiar with the second half of Apollonius' treatise, which by then had not yet been translated from the Arabic into Latin, as far as is known. Apollonius knew astronomy and Ptolemy undoubtedly knew conics, but they did not use conics as celestial orbits. Kepler did so, and the mathematics which he pre-required should have been readily available to Apollonius and Ptolemy. Therefore, from a certain pragmatic approach, Kepler was a direct successor to Apollonius and Ptolemy.

The Isaac Newton of the *Principia* is by the principal subject matter of the treatise a direct successor to Kepler and by the manner and style of its conceptualizations a direct successor to Archimedes. What makes the *Principia* the book that it is are not its Definitions and its Laws and its views on Time and Space and not even its Central Forces if taken by themselves and on their own merit. The *Principia* is what it is because with this congeries of definitions, laws, concepts of time and space, and his gravitational force, and by mathematically controlled reasoning, Newton constructed and deduced what Kepler had only divined and posited, and because into this design of Kepler's celestial orbits, and from the identical blueprints, Newton also fitted the (then) earthbound trajectories of Galileo and Toricelli. The 20th century has apparently verified that even artificial celestial bodies conform to the Newtonian theory of the 17th century, and, what is most remarkable, nobody in the 20th century doubted that they would (except of course for "relativistic" adjustments if such should be called for, eventually).

Now, outwardly the *Principia* is written in a pure idiom of Archimedes and Apollonius, and outwardly there is hardly any mathematics invoked or presupposed which should not have been quickly accessible to them. Newton was very familiar with Descartes' work, and in his book on plane curves Newton masters the method of Descartes in analytical detail much better than Descartes himself. Also Newton was quite skilled in the use of symbols and functions from the beginnings of his mathematical career, as his early work on power series and also on fluxions shows. But in the *Principia* no attempt is made to modify the Greek idiom by the use of Descartes' innovations or to employ symbols and functions analytically; even the inviting trait of discursiveness that animates the *Opticks* is hardly present in the *Principia*. Furthermore, and this we would like to stress, Newton's definitions of limit and derivative ("ultimate ratio") as presented in the *Principia* seem to us to be such, in syntax and in content, that a personal disciple to Archimedes should have been able to compose them. At any rate, it would be a major task to substantiate philosophically that a student of Archimedes could not possibly have arrived at them because of their being removed from and alien to the Hellenic-Hellenistic milieu of natural philosophy within which the developments from Eudoxus to Archimedes and Apollonius had taken place.

However, inwardly, the Euclidean Space that underlies the *Principia* is mathematically not quite the same as the Euclidean Space that underlies Greek mathematics (and physics) from Thales to Apollonius. The geometry of the

Greeks emphasized congruences and similarities between figures, that is, in analytical parlance, orthogonal and homothetic transformations of the underlying space. The Greeks themselves were half-aware of this. Late Greek traditions about the origin of their mathematics, whether true or legendary, ascribe to Thales mathematical exploits which involve congruences and similarities emphatically. The Euclidean Space of the *Principia* continued to be all this, but it was also something new in addition. Several significant physical entities of the *Principia,* namely, velocities, momenta, and forces, are, by mathematical structure, vectors, that is, elements of vector fields, and vectorial composition and decomposition of these entities constitute an innermost scheme of the entire theory. This means that the mathematical space of the *Principia,* in addition to being the Greek Euclidean substratum, also carries a so-called affine structure, in the sense that with each point of the space there is associated a three-dimensional vector space over real coefficients, and that parallelism and equality between vectors which emanate from different points are also envisaged. In the course of the 18th century the vectorial statements of Newton and others were gradually transferred and reinterpreted into analytical statements. This involved Euclidean coordinates and eventually also "generalized" coordinates of Lagrange. In these analytical developments the original vectorial character of the space of Newton ceased to be featured. But in the 19th century, partly within the concepts of Hamilton's theory of quaternions, the underlying vectorial trait was gradually recalled, and even taken account of by a symbolism of its own which has remained in use. The 20th century widened the concept of a vector into the broader concept of a tensor. This called forth readjustments in mechanics and physics and affected pure mathematics even more profoundly.

Newton asserts the vectorial character of mechanical

force in "Corollary II to the second law" in Book I of the *Principia,* and his assertion consists of two parts. The first part states that any two forces give rise to a resultant force which is the vectorial composition of the two. The second part is in the nature of a converse to the first part, and it states that in a plane a force can be decomposed into two component forces along any two non-parallel lines. The second part of the assertion is much more recondite than the first, even if applied to the simpler case of velocities rather than forces. It is a simple observational fact that if two persons pull at a load in different directions then the load moves along some suitable diagonal, and this fact was undoubtedly known to Aristotle as commonly asserted. But it is an entirely different assertion to say that if in a plane a force is given "concretely" in a well-determined line of action, then purely speculatively it can be imagined to be a vectorial sum of two "fictitious" component forces along "fictitious" lines of action of arbitrary choice; and there is nothing to suggest that such a possibility was in the minds of natural philosophers before the 17th century, say.

3. THE 18TH CENTURY

The 18th century was an age of organization, for mechanics and even for mathematics. It produced several so-called Principles of mechanics and some special theories and methods. Books from the 18th century are frequently still readable, and many of their notations have maintained themselves into current or relatively recent use. The organizing activity was not all spent on a continuation and elaboration of what the 17th century had achieved; a large part of it was spent on a re-creation and reestablishment of the subject matter from its roots.

The 18th century introduced, more or less formally, the concept of a mass particle in the sense of a geometric mass point, and of a finite system of mass particles with general forces and constraints acting between them, and it recognized that the theory of such finite systems is a guide and pilot theory for rational mechanics as a whole. But it also studied extensively the so-called mechanics of continua, which comprises large and detailed theories of its own, chief among which are hydrodynamics, acoustics, and the general theory of elasticity. Acoustics on a very elementary level is as old as the Pythagoreans, a hydrostatical theorem is one of Archimedes' great claims to fame, and rudimentary beginnings of elasticity have been ascribed to the Stoics.

The organization of mechanics necessitated the creation of a considerable amount of mathematics, mainly in Analysis. Of this mathematics, the calculus of variation was instigated largely by mechanics of particles, but other theories were instigated largely by mechanics of continua. Virtually all of partial differential equations started in this way, especially the mathematical theory of waves, which eventually became the hallmark of theoretical physics in all its parts. Mechanics of continua also created half of Fourier analysis. The other half was created by the theory of heat, which also contributed the name of the theory. Mechanics of continua also had a fair share in the promotion of potential theory. But it should be said that the concept of potential energy originated in the Lagrangean theory of finite systems of particles, and that, on the other hand, through the work of Poisson and others, the concept of potential acquired the status of a commanding physical entity not within mechanics but within electrostatics.

Finally, it appears that mechanics of continua had a share in the emergence of tensor theory, which after slow stages of nascency began to take shape towards the end of the 19th century and in the 20th century has been penetrating various areas of physics and mathematics.

Important entities in the mechanics of continua, at whose mathematical conceptualization the emergent theory of tensors was aiming, were what are nowadays the pressure and stress tensor and the accompanying deformation tensor. Pressures, tensions, stresses, and suchlike entities are species of "force," but phenomenologically they seem not to subordinate themselves to the concept of force as introduced in Book I of the *Principia,* the mathematical substratum of which is a vector of three components issuing at a point. In fact they seem to be entities which associate themselves with the various surface elements through a point in space rather than with a point of space by itself. In the outgoing 18th and beginning 19th centuries efforts were made to overcome these disparities by reducing all surface forces to purely vectorial forces through appropriate molecular approximations; but there were also parallel efforts to conceptualize things as they present themselves phenomenologically. In this conceptualization there are two types of force, point forces and surface forces, and they are linked by certain basic laws which are modeled on Newton–Lagrange prototypes. A surface force is a "stress," as introduced in note 1, above. As there stated, pressure is a particular kind of stress, and although it is a one-component quantity, yet as a mathematical object it is far removed from anything which Archimedes might have been able to conceptualize in his work on floating bodies against the background of the mathematics then alive.

A particular situation arises in the case of rigid bodies. They are continuous media, but tensorial forces need not be envisaged, although a tensorial setting does illumine the context. The theory of rigid bodies is a doctrine by itself. Large parts of it are geometrical, but it is a "modern" geometry of analytical variability, and not of Greek stationarity. The theory was fully started in the 18th century by L. Euler and is a showpiece among the many masterpieces of this great master. In retrospect, his kinematics is a first meaningful study of groups, namely, of one-dimensional continuous groups of orthogonal transformations in Euclidean space, well over 100 years before Sophus Lie, Erlanger Programs, and the like.[2] His dynamics was a transfer, with suitable elaborations, of appropriate statements for finite systems of points to the continuous case, without a trace of brooding over the melancholy fact that there are no truly rigid bodies in nature.

All told, the 18th century achieved very much for the theory of continuous media and the mathematics involved in it. And yet, the historical "image" of 18th-century mechanics is composed of achievements which relate to finite systems of particles primarily.

Of great originality are two so-called differential principles, the principle of Bernoulli and the principle of d'Alembert. They can be stated and appreciated separately; but what has given them their status is the fact that jointly they are the foundation for the general Lagrange equations of motions for finite systems of particles. In any rationalization of these equations the two principles are involved deeply and inseparably.

The principle of d'Alembert mathematically "transforms" a dynamic system of particles into a static one, and it views the so-called Newtonian equations of motion as if they were conditions for equilibrium. If a single point moves on a straight line under the influence of an

[2] The so-called Erlanger Program of Felix Klein is an essay of 1872 with numerous later additions (for complete texts see Klein's *Gesammelte mathematische Abhandlungen,* Vol. 1, Berlin, 1921, pp. 411–612; an English translation of the original essay is in *Bulletin of the New York Mathematical Society,* Vol. 2, 1899, pp. 215–249), which envisages the significance of groups of transformations for the geometric fabric of spaces. The systematic theory of groups of transformations of Sophus Lie, which is a leading bequest of 19th century to the 20th, was also begun around 1870.

"external" or "impressed" force F, then the principle of d'Alembert puts the equation of motion in the form

$$-m \frac{d^2x}{dt^2} + F = 0$$

and, as many in the 19th century understood the principle, it "interprets" the term

$$-m \frac{d^2x}{dt^2}$$

as representing also a force, the so-called inertial force. Thus it interprets the equation of motion to mean that the "total" force, that is, the sum of external force and of inertial force, is zero. In a general system of mass points it asserts that suitable vectorial sums of external forces and of inertial forces are each zero. Since the notion of an external force is usually an "abstraction," for instance if it is a gravitational force, the inertial force of d'Alembert is in a sense an abstraction from an abstraction, or rather an abstraction vertically·imposed on a previous one. Such "multifold" abstractions, that is, abstractions from abstractions, abstractions from abstractions from abstractions, etc., if performed in cognitive depth, generate much of the motive power of modern mathematics and also modern science; and, as we have stated before, not many such will be found among all the "ideations" from Plato to Plotinus to Abélard to Ockham, especially not from Plato to Plotinus.

The principle of Bernoulli is more technical, and in a sense even more daring. For our purposes we will present it in two stages. The first stage will be seemingly a mathematical tautology in preparation of the second stage. But we should point out that we will follow the presentation of the principle by Lagrange rather than Bernoulli's own.

If in a finite system of mass points we denote the totality

of all components of all mass points in a uniform lettering by

$$x_1, x_2, \ldots, x_N \tag{1}$$

and the corresponding components of the external forces by

$$F_1, F_2, \ldots, F_N$$

then in the case of equilibrium all these components of force are zero. That is, we have

$$F_1 = 0, F_2 = 0, \ldots, F_N = 0 \tag{2}$$

Purely mathematically this set of N conditions can be replaced by the one condition

$$F_1 \cdot \lambda_1 + F_2 \cdot \lambda_2 + \ldots + F_N \cdot \lambda_N = 0 \tag{3}$$

which is assumed to hold for all possible choices of real numbers

$$\lambda_1, \lambda_2, \ldots, \lambda_N$$

Now the principle interprets each number λ_n as a "virtual" displacement of the corresponding space component x_n. Such a displacement is usually denoted by δx_n, and thus condition (3) is now written as

$$F_1 \cdot \delta x_1 + F_2 \cdot \delta x_2 + \ldots + F_N \cdot \delta x_N = 0 \tag{4}$$

The first interpretative gain of this formulation was as follows. The expression on the left side, that is, the sum

$$F_1 \cdot \delta x_1 + F_2 \cdot \delta x_2 + \ldots + F_N \cdot \delta x_N \tag{5}$$

was conceived as a certain "action" of the given forces along the given virtual displacement—which the 19th century eventually called "work"—and equation (4) is now interpreted to state that the system of forces is in equilibrium if and only if this action is 0 along any virtual displacement whatsoever.

This is the first stage of the Bernoulli principle. For the second stage we reconsider matters from the following approach. Thus far we have tacitly assumed that our system is a "free" system of particles, that is, that the presence of external forces is all that influences the motion of the particles. But now we take a system of particles in which so-called constraints are present, where constraints are certain primary conditions which prescribe that only certain geometrical configurations of the mass points can occur. Also the constraints are assumed to be independent of the external forces and to be superimposed on these. Now, the second stage of the Bernoulli principle states that for a system with constraints the necessary and sufficient conditions for equilibrium continue to be the same as in a free system. Thus, it is again equation (4), and $F_1, \ldots,$ F_N are again the components of external forces, and $\delta x_1, \ldots, \delta x_N$ are again virtual displacements. But, and this is decisive, *only such virtual displacements are to be considered as are compatible with the constraints.*

It took a long time before it was generally understood that the principle of Bernoulli if applied to systems with constraints is a far-reaching hypothesis of axiomatic nature. L. Poinsot in an angry note in 1838 (*Journal de Mathématiques pures et appliquées,* Vol. 3, pp. 244–248) castigated Laplace for misrepresenting this point in his *Mécanique céleste.* And C. G. Jacobi, when stating matters very clearly in his *Vorlesungen über Dynamik* (1842–43, published 1865), still deemed it appropriate to invoke the great authority of Gauss in support of his statement. (After the death of Gauss it was not possible to verify from his papers that such an assent to a view of Jacobi had been given.)

Lagrange obtained his equations of motion in the following manner. He superimposed the principle of d'Alembert on the principle of Bernoulli; that is, he applied the second stage of the principle of Bernoulli to forces which are sums of external and inertial forces. Then, by a purely mathematical argument, which is one of his leading mathematical achievements, he transformed the constraining *conditions* also into constraining *forces*. As a result, his equations of motion for a finite system with constraints have the same analytical form as for a system without constraints, except that in addition to the external forces there are some further "ideal" forces which account for the constraints. These ideal forces are perpendicular to those virtual displacements which are compatible with the constraints so that they make no contribution to expression (5). Also, the actual sizes of the constraining forces are measured by the values of so-called Lagrange multipliers.

This was only half of what Lagrange did in this area, and the second half is the one that really matters. Lagrange subjects the original space coordinates (1) to a general analytic transformation of variables. This gives rise to new variables, again of the same total number N, and they are frequently denoted by the letters

$$q_1, q_2, \ldots, q_N \tag{6}$$

These are his so-called general coordinates, and Lagrange translates the previous equations of motion into new equations for the new coordinates. Now, if there are constraints in the system, and if they are expressible by s "independent" equations between the original N coordinates (1) then it is possible to choose the general coordinates (6) in such a way that the constraining conditions become the requirement that s among the new coordinates shall be zero, say

$$q_{N-s+1} = 0, \, q_{N-s+2} = 0, \ldots, q_N = 0 \tag{7}$$

The remaining $N-s$ coordinates

$$q_1, q_2, \ldots, q_{N-s} \tag{8}$$

are then unconstrained, and these are Lagrangean "free parameters" of the system. If general parameters have been so chosen, then the system of equations of motion decomposes into two parts, a primary part and a secondary part. The primary part involves only the free parameters (8), does *not* make express reference to constraining forces, and completely determines the solutions of the problem. The secondary part, on the other hand, only determines the constraining forces and nothing else. Thus the secondary part may be ignored altogether, and textbooks usually omit it.

Constraints and constraining forces, and attempts to understand their nature and effects, are as old as mechanics itself and a full history of the genesis of these concepts would be virtually coextensive with a history of the genesis of mechanics itself, at least of the mechanics of weights and ordinary bodies. One would certainly have to take into account the work of Jordanus (13th century), and perhaps even go back to Heron of Alexandria. Stevin and Galileo operate with constraints and constraining forces in the open, and Newton does so even more expressly. And yet their analytical formulations in the 18th century, apparently more-or-less due to Lagrange, constituted a major novelty nonetheless.

The emergence of general coordinates and free parameters in the 18th century was intellectually a prelude to the rise of multi-dimensional geometries, Euclidean and non-Euclidean, in the 19th century. Some of the creators of these geometries were not only experts in rational mechanics but even leaders in it. We think that, by evolution of ideas, the geometries of the 19th century were, in good part, descendant from 18th-century mechanics, and not only in the general sense that any period of mathematics is descendant from the preceding one. The total development was an evolutionary process. In the 17th and 18th centuries the structure of space and place was Euclidean as a matter of course. In the 19th century the possibilities of non-Euclidean structure were being explored for geometry and for geometry alone. By the end of the century, when the mathematics was ready for it, first steps were taken towards inserting non-Euclidean features into mechanics also. And in the course of the 20th century some non-Euclidean (that is "relativistic") features of mechanics have already become basic and commonplace.

Some philosophers and mathematicians unduly single out the emergence of non-Euclidean geometry as the decisive mathematical event of the 19th century. Non-Euclidean geometry is all-important for any confrontation between modern mathematics and Greek mathematics. But it was not the only great achievement of the 19th century, far from it; and against the background of what the 18th century had been preparing for, it was not a startling innovation but rather more a "natural" event, albeit a very notable one. Even if mathematicians and philosophers at large were aroused and profoundly affected by the works of J. Bolyai (1802–1860) and N. I. Lobachevsky (1793–1856) about non-Euclidean geometry, which became known in the decade 1831–1840, yet mathematicians like Jacobi, who had been steeped in general mechanics and habituated to generalized coordinates, did not show surprise at these developments; and there is nothing to suggest to me that mathematicians of this level had the feeling that a novel era of natural philosophy was breaking in on them.

It was providential that the 17th and 18th centuries, especially the 17th, took the Euclidean structure of space for granted and knew of no other. Newton did find the law of gravitation for Euclidean space. But if, by some inspiration, he had known about non-Euclidean possibilities and had tried to find the law of gravitation directly for a space with curvature, he might have been led woefully astray. Greek mathematics, viewed in retrospect, undertook to develop things out of order and came to grief. Instead of developing a concept of real number as a common mathematical measure for physical quantities, Greek mathematics undertook to develop directly a mathematical theory of general physical quantities. They seemingly succeeded, as the fifth book of Euclid and the monumental works of Archimedes and Apollonius testify. But the resulting mathematics was too inflexible to adjust itself to needs of mechanics and physics, and too cumbersome and heavyweight to propel even itself far beyond the platform on which it had gained its crown of laurels.

In the evolution of modern mathematics and mechanics a contingent disharmonious feature was the timing of the rise of the great Kantian philosophical system. Having come into being in the 18th century, it tied itself to Euclidism, and in fact to a doctrinaire 17th-century version of it, and adherents in later generations wished it had not done so.

4. THE PRINCIPLE OF LEAST ACTION

A lively activity in the field of 18th-century mechanics and mathematics was the search for a suitable version of the principle of least action. The search was beset with difficulties and misunderstandings because, as seen in retrospect, it involved the search for the concepts of both kinetic and potential energy on a level of quantitative precision. These concepts, however, were apparently difficult and recondite, and, in fact, only the 19th century was equipped to master them.

The search for the concept of energy began early. A notion of energy, or perhaps a set of companion notions, was a dominant and ever recurrent feature of Aristotle's philosophical system. One may even claim for him that the distinction between kinetic and potential energy was already foreshadowed in his work, especially in his famous statement that "motion (that is the Aristotelean *kinesis*) is the actualization of that which is potentially, as such." But in cold fact it must be said that there is a long distance between his notion of energy and a concept, or even preconcept, of energy that fits into mechanics as we understand it. Some of his medieval followers may have come closer to it, but the important researches of Anneliese Maier suggest that the Schoolmen had great difficulties in keeping apart force, energy, and momentum.

Almost everybody in the 17th and 18th centuries had something to say about *vis viva,* that is kinetic energy, except Isaac Newton, who most prudently avoided getting involved in it, perhaps because, unlike force, energy could not be spatially exhibited. Huygens, on the other hand, was keenly interested in it and made an important application of it. He reasoned that, in an elastic collision of two masses, not only that total translation momentum remains continuous throughout the collision, which fact had already been known before, but the total kinetic energy too.

Lagrange was the first to introduce potential energy. He also gave a satisfactory version of the principle of least action, and in it he also uses the concept of total energy as the sum of kinetic and potential energy. The Lagrange version of the principle has been overshadowed by the Hamiltonian principle, but Lagrange can claim credit for the fact that his version, which he found much before Hamilton, is in basic mathematical structure the more difficult one.

Although Lagrange introduced potential energy and applied it, the theory of it was elaborated by others (Laplace, Poisson, Gauss, Green, etc.). Also the actual interest in potential energy came not from mechanics but from electrostatics, in which it is the identifying concept, almost. Similarly the concept of total energy of a system, although originating in mechanics, acquired its status in the theory of heat, in which, again, it is the dominant concept.

Theoretical Physics, as against rational mechanics, fully began only with the concept of energy, after mechanics had completed its main phase without yet concentrating on it. The (prequantum) energy was measured by arbitrary real numbers, and it is perhaps more than a coincidence that the proper conceptualizations of real numbers and of energy took place virtually simultaneously in the 19th century.

Simple cases of the Lagrange version of the principle of least action are rarely presented in textbooks in analytical detail and we are going to append one.

Let T denote the kinetic energy of a conservative system, U its potential energy, and let t denote time. The principle states that the equations of motion are obtained by the variation

$$\delta' \int_{t_0}^{t_1} 2T \, dt = 0 \qquad (9)$$

under the side condition

$$\delta' (T + U) = 0 \qquad (10)$$

where δ' denotes the variation not only of the space variables but also of the time scale, both between fixed endpoints.

Consider the case of a mass point m moving on the x-axis. Its kinetic energy is

$$T = \frac{m}{2}\left(\frac{dx}{dt}\right)^2$$

and we have to vary the integral

$$m \int_{t_0}^{t_1} \left(\frac{dx}{dt}\right)^2 dt \qquad (11)$$

Since t is also to be varied we must introduce a function $t = t(u)$ over an interval $u_0 \leqq u \leqq u_1$. Expression (11) is then

$$m \int_{u_0}^{u_1} \left(\frac{dx}{du}\right)^2 \left(\frac{dt}{du}\right)^{-1} du \qquad (12)$$

We must now replace x by $x + \epsilon\xi$ and t by $t + \epsilon\tau$, where ξ and τ are finite displacements and ϵ is a small real parameter.

If we substitute this in (12) we obtain a function

$$A(\epsilon) = m \int_u^{u_1} \left(\frac{dx}{du} + \epsilon \frac{d\xi}{du}\right)^2 \left(\frac{dt}{du} + \epsilon \frac{d\tau}{du}\right)^{-1} du \quad (13)$$

and the requirement (9) is

$$\left.\frac{dA(\epsilon)}{d\epsilon}\right|_{\epsilon=0} = 0$$

By carrying out the differentiation in (13) and then putting $\epsilon = 0$ we obtain

$$2m \int_{u_0}^{u_1} \frac{dx}{du} \frac{d\xi}{du} \left(\frac{dt}{du}\right)^{-1} du - m \int_{u_0}^{u_1} \left(\frac{dx}{du}\right)^2 \left(\frac{dt}{du}\right)^{-2} \frac{d\tau}{du} du = 0$$

$$(14)$$

Next, for the requirement (10) we have to introduce the function

$$B(\epsilon) = \frac{m}{2}\left(\frac{dx}{du} + \epsilon \frac{d\xi}{du}\right)^2 \left(\frac{dt}{du} + \epsilon \frac{d\tau}{du}\right)^{-2} + U(x + \epsilon\xi)$$

and the requirement itself is

$$\left.\frac{dB(\epsilon)}{d\epsilon}\right|_{\epsilon=0} = 0$$

By carrying out the differentiation and putting $\epsilon = 0$ we obtain

$$m \frac{dx}{du} \cdot \frac{d\xi}{du}\left(\frac{dt}{du}\right)^{-2} - m\left(\frac{dx}{du}\right)^2 \left(\frac{dt}{du}\right)^{-3}\frac{d\tau}{du} + \frac{\partial U}{\partial x}\xi = 0$$

From this we obtain

$$m\left(\frac{dx}{du}\right)^2 \left(\frac{dt}{du}\right)^{-2}\frac{d\tau}{du} = m \frac{dx}{du}\frac{d\xi}{du}\left(\frac{dt}{du}\right)^{-1} + \frac{\partial U}{\partial x}\xi \frac{dt}{du}$$

and if we substitute this into (14) the result is

$$\int_{u_o}^{u_1} \left[m \frac{dx}{du} \frac{d\xi}{du} \left(\frac{dt}{du} \right)^{-1} - \frac{\partial U}{\partial x} \xi \frac{dt}{du} \right] du = 0 \qquad (15)$$

Thus, the time displacement τ has been eliminated and we can now return from the variable u to the original variable t. Then, (15) becomes

$$\int_{t_0}^{t_1} \left[m \frac{dx}{dt} \frac{d\xi}{dt} - \frac{\partial U}{\partial x} \xi \right] dt = 0 \qquad (16)$$

Now, by partial integration we have

$$\int_{t_0}^{t_1} \frac{dx}{dt} \frac{d\xi}{dt} dt = \frac{dx}{dt} \xi \Big|_{t_0}^{t_1} - \int_{t_0}^{t_1} \frac{d^2x}{dt^2} \xi \, dt \qquad (17)$$

Next, we assume that the displacement ξ is 0 at the time endpoints t_0, t_1, and we substitute (17) into (16). The result is, after changing the sign,

$$\int_{t_0}^{t_1} \left[m \frac{d^2x}{dt^2} + \frac{\partial U}{\partial x} \right] \xi \, dt = 0$$

Now, for a given function $x(t)$ between the limits t_0, t_1 this must hold for any displacement $\xi(t)$ which is 0 at t_0 and t_1. This can only occur if the factor of ξ under the integral is 0, that is,

$$m \frac{d^2x}{dt^2} + \frac{\partial U}{\partial x} = 0$$

But this is indeed the equation of motion of a mass point m on the x-axis subject to a force which is derived from a potential $U(x)$.[3]

[3] For the general case of Lagrange's own version of the principle of least action (but with much less analytical detail), see for instance an article by L. Nordheim in *Handbuch der Physik*, edited by H. Geiger and Karl Scheel, Vol. 5 (Berlin, 1927), pp. 83ff; or, an article by J. L. Synge, in the second edition of this *Handbuch*, also called *Encyclopedia of Physics*, edited by S. Flügge, Vol. III/1 (Berlin, 1960), pp. 136ff. Most handbooks name this version after Euler and Maupertuis.

CHAPTER 6

THE SIGNIFICANCE OF
SOME BASIC
MATHEMATICAL CONCEPTIONS
FOR PHYSICS *

THE PRESENT chapter supplements and continues the preceding one, and we will again make observations on the relations between mathematics and physics on the level of abstractions and conceptualizations; in the preceding chapter we emphasized mechanics, and at present we will be more concerned with physics in general.

The Greeks created rational philosophy and a general conception of rational philosophy, and they created a general conception of mathematics and a general conception of physics. Their mathematics is justly renowned; the Greeks imparted to it from the beginning a certain metaphysical accent which has become a distinguishing mark of mathematics in general. They developed an astronomy and geometrical optics which were quite mathematical, and the Greeks themselves wondered whether, and in what sense, these could be distinguished from mathematics proper.

Yet if one undertakes to assess in retrospection and pragmatically the outcome of Greek mathematics and physics, and if one confronts the mathematics and physics of Greek antiquity with modern mathematics and physics, then one may nevertheless find that Greek physics as a whole, and in its inward structure, never developed into a "theoretical" system, that the system of Greek mathe-

* Originally in *Isis,* Vol. 54 (1963), pp. 179–205. Revised.

matics had severe limitations and shortcomings, and that these inadequacies made Greek mathematics unsuited to promote the rise of theoretical physics as we have it or to mold scientific thinking in our sense. A mathematics that was much more suited to do this began only after the Middle Ages, and its first impact on physics and other science was to promote the rise of a Rational Mechanics, which held the stage of science during the 17th and 18th centuries and without which the general mathematization of physics and then, progressively, of ever larger areas of so-called exact science and also of other (even social) science would not have come about. In what follows, some features of this mathematics, and their involvement with physics will be presented.

I. THE SIGNIFICANCE OF MULTIPLICATION

1. GREEK MAGNITUDES

From whatever reasons, Greek mathematics did not create a conception of real numbers. Instead, it was satisfied with, and became arrested on the precarious notion of magnitude ($= \mu\acute{\epsilon}\gamma\epsilon\theta$os), which the Greeks never defined, in or out of mathematics, but which was their substitute for real number. For instance, lengths were one kind of magnitude, areas were another kind of magnitude, weights were still another kind of magnitude, etc. The Greeks could envisage the ratio $P_1 : P_2$ for any two values of the same magnitude, and the ratio $L_1 : L_2$ for any two values of any other magnitude, and they could envisage the proportion

$$P_1 : P_2 = L_1 : L_2 \qquad (1)$$

between them; they could verify, by a famous criterion which was a summit achievement of Greek mathematics, whether the proportion does or does not hold for the four specific values given. They also had a calculus of ratios and proportions with which they could operate competently and skilfully. But the calculus was restrictive and stultifying, it did not have a dynamism of evolution, and even Archimedes could not break out of its constrictions. The fact is that the "classical" Greek mathematics of Euclid, Archimedes, and Apollonius could never form the conceptual product

$$P \cdot L$$

for two magnitudes P and L in general. In the very special case in which P and L were both lengths, a product did "exist" for them, and it was an area; similarly if one of the factors was a length and the other an area, a product did exist and it was a volume. But these exceptions were made unconsciously and unreflectively; nobody "philosophized" about them or in any way suggested to extend the product to other magnitudes. And, as we have already stated in Chapter 5, in consequence of this limitation, the formation of a momentum in mechanics was retarded by 2,000 years.

Before becoming too impatient with the Greeks for not having created a product ab of real numbers a and b, one ought to consider that the actual conceptualization-*cum*-symbolization of such a product did not come much earlier than the creation of a group product ab for elements a and b of a general group G, and that without such a general group product there would be no avant garde physics of our today. In other words, the creation of the product ab for real numbers was not a belated terminal phase to the development of the Greeks, but a preparatory initial phase for the developments of today.

2. PHYSICAL DIMENSIONS

The product $P \cdot L$ not only multiplies the numerical values of P and L but it also forms a new "magnitude" out of the two magnitudes P and L. The 19th century extended this process of multiplication from magnitudes in mechanics to magnitudes in all of physics, especially to those which originate in electricity and magnetism; in fact it created a certain doctrine of "physical dimensions," where "physical dimension" is the same as "physical magnitude" except that no quantitative value is attached to it. J. Clerk Maxwell and Fleeming Jenkin introduced in 1863 [1] the symbol $[P]$ for the "dimension" of any magnitude P, whether the magnitude be "simple" or "composite," and they stipulated expressly that in a composite dimension

$$[P_1^{a_1} \cdot P_2^{a_2} \ldots P_k^{a_k}] \tag{2}$$

the exponents $a_1, a_2, \ldots, a_k$ may be arbitrary real numbers. In 1822, in a short but systematic essay on physical dimension, which had been given by J. Fourier in Chapter II, Section 9, of the *Theorie de la chaleur,* the exponents had still been presumed to be integers only, but they already were positive or negative integers without further restriction.

For instance, if $M, L,$ and T are mass, length, and time, then

$$[\text{energy}] = [\text{work}] = [M \cdot L^2 \cdot T^{-2}]$$

and

$$[\text{"action"}] = [M \cdot L^2 \cdot T^{-1}] \tag{3}$$

We mention that the absolute constant h of quantum theory has the dimension (3) of an "action," and that Max Planck (1858–1947), the creator of the constant (1900), was in the habit of pointing this out with some emphasis.

The possibility of a twin conceptualization of a "physical dimension" and of a formal product (2) by which to symbolize it is no trifling matter. Not even the separate notion of a "physical dimension" by itself, without its involvement with a symbolism, will be found anywhere in Greek philosophy of nature, although the total course of

[1] It was published ten years later in *Reports of the Committee on Electrical Standards, British Association for the Advancement of Science* (1873), pp. 59ff.

this philosophy extended over more than a thousand years, namely from Thales (585 B.C.) to Proclus (5th century A.D.) and Simplicius (6th century A.D.) and included Heraclitus, Parmenides, Plato, Democritus, and Aristotle; and only some blurred and passing hints of it might be detected in utterances of some of the most forward-thinking philosophers in the Middle Ages.

The over-all intellectual development within which the twin conceptualization came about is a certain comprehensive process of algebraization and symbolization of mathematics as a whole, and also of rational thinking in general. The process began in the Middle Ages and in a sense is still continuing; a particular manifestation of it is the gradual rise of algebra proper as a main division of mathematics and perhaps also the emergence of a symbolic logic in the 19th century.

3. PROPORTIONS

A peculiar application of the Greek calculus of proportions occurs in the *Timaeus* of Plato.[2] If a cube has a side of length c, then its volume is c^3. The Greeks would state this by saying, in our notation, that if two cubes have sides a, b, then the ratio of their volumes is the triplicate of the ratio $a : b$; which means that the quantities a^3, b^3 are the extreme terms in the chain proportion

$$a^3 : a^2b = a^2b : ab^2 = ab^2 : b^3$$

This proportion contains four terms, namely

$$a^3, \ a^2b, \ ab^2, \ b^3$$

and therefore, according to a Pythagorean view,[3] spatiality is associated with the enumerative role of the integer 4.

[2] To the following, see the commentary of F. M. Cornford, *Plato's Cosmology* (1937), pp. 43ff.

[3] See, for instance, G. S. Kirk and J. E. Raven, *The Presocratic Philosophers* (1957), pp. 253–254. Also Aristotle, *Metaphysica*, 1090 b 20–25.

Now, the *Timaeus* asserts (31 B–34 A) that this integer 4 also enumerates the number of space-filling elements of the cosmos, which are earth, water, air, and fire. It is tempting to disparage a rationalization of this kind; but before yielding to such a temptation, one ought to consider that contemporary combinatorial topology shares the Pythagorean view that a three-dimensional simplex (that is tetrahedron) is determined by its four vertices, and we note that the physical and cosmological work of Arthur Eddington (1882–1944) also had a numerological slant.[4]

4. THE FUNDAMENTAL THEOREM OF PRIME NUMBERS

The Greek "horror of multiplication" also left a mark on the arithmetical theory of prime numbers. The ninth book of Euclid's *Elements* (300 B.C.) is a celebrated treatise on the theory of numbers, mainly prime numbers. But it does not have the so-called fundamental theorem which states that every integer can be represented as a finite product of prime numbers in which some factors may repeat themselves, and that the representation is unique. The product of equal factors is a power p^n, and therefore the theorem states in symbols that every integer N has a unique representation as a product

$$p_1^{n_1} \cdot p_2^{n_2} \cdots p_k^{n_k} \tag{4}$$

in which $p_1, p_2, \ldots, p_k$ are different prime numbers, say

$$p_1 < p_2 < \cdots < p_k$$

and the exponents $n_1, n_2, \ldots, n_k$ are some integers. What makes the omission of the fundamental theorem from Euclid's treatise noticeable is the fact that Euclid presents

[4] See, for instance, his posthumous work, *Fundamental Theory* (1946).

other theorems which, in modern accounts, are syllogistically very close to it and are deemed to be auxiliary to it.

We do not assert that Euclid and the later Greeks were not able to "find" the fundamental theorem simply because it involves the formation of products. In fact, in the case of numbers the Greeks did form products, of two or more factors, even if the number of factors was unspecified. But, to Greek mathematics our theorem would have offered another very serious difficulty; and this difficulty was interlocked with the difficulty of formation of products in such a manner and to such an extent that the overcoming of this difficulty would have taxed the Greek capacity greatly. The fundamental theorem is only an existence theorem. It only asserts the existence of a representation (4) and it is not concerned with the question of finding or constructing the integers p_1, p_2, . . . , p_k and n_1, n_2, . . . , n_k if N is given. At first sight one might think that this fact would make the theorem easier and less complicated, but actually to a Greek this would have made the theorem only harder to find. The Greeks never quite arrived at the insight that in mathematics it might be *conceivable* to have an "existency" which is *totally* divorced from "constructibility"; and the fundamental theorem is an existence theorem not from choice but from necessity. There are no ready-made effective formulas, beyond the sieve of Eratosthenes, for actually constructing or finding the product (4) if N is some general integer; that is, for given N the number k of monomials in (4) is not only unspecified but incapable of being specified by a formula in the ordinary sense, and the various prime factors p_1, . . . , p_k with their exponents n_1, . . . , n_k are all unspecifiable and unpredictable in the same sense. But to form a product of many factors with such undetermined and unspecified data as this situation demands would have been much more than a Greek mathematician might have been expected to accomplish.[5]

In fact, the fundamental theorem of prime numbers, that is, the assertion that any integer N can be (uniquely) represented as a product (4) of prime numbers, not only eluded Euclid but it continued to elude mathematicians for another 2,100 years, and it eluded all the great mathematicians of the 17th and 18th centuries as well. The first time the theorem became generally known was in 1801, when it was presented in print by C. F. Gauss in § 16 of his *Disquisitiones arithmeticae*.

Arithmeticians have been showing embarrassment over the fact that the express formulation of the theorem came so late, and they have even been trying to "pre-date" it. But there is no substance to assertions that the fundamental theorem had been consciously known to mathematicians before Gauss, but that they had neglected to make the fact known. We think that the 17th, and even the 18th, century was not yet ready for the peculiar kind of mathematical abstraction which the "fundamental theorem" involves; just as only the 19th century was comfortably prepared to comceptualize satisfactorily the notions of real number, limit, derivative, convergence, etc.

II. THE SIGNIFICANCE OF FUNCTIONS

5. THE RISE OF FUNCTIONS

A mathematical object of paramount significance to mathematics and science is the notion of "function." There are two main descriptions of the notion of function, one in terms of "correspondences" and another in terms of

[5] The gist of the above "explanation" for the absence of the "fundamental theorem" from Euclid's treatise was communicated orally by the present author to G. H. Hardy in 1933, as noted in the book of G. H. Hardy and E. M. Wright, *An Introduction to the Theory of Numbers* (1938), Chap. XII, §12.5, p. 181 and a note to it on p. 188.

"relations," and they are both illuminating. But in reality the notion of function is not definable, and any would-be definition is tautological.[6]

Functions are a distinguishing attribute of modern mathematics, perhaps the most profoundly distinguishing of all. Functions began to emerge in the Middle Ages, and one frequently associates the name of Nicole Oresme (14th century) with the first clear symptoms of their emergence. In its innermost structure Greek mathematics was a mathematics entirely without functions and without any orientation towards functions; and whatever rudiments of functions there may have been in Later Greek mathematics could probably be traced to its merger with Babylonian mathematics in which something in the nature of functions was trying to break through.

It has been rightly emphasized by Oswald Spengler [7] that the presence of functions in mathematics signifies a certain all-around mobility and their absence a certain all-around stationarity, as evidenced by contrasts between Greek and modern civilizations in some of their essential aspects.

As far as physics is concerned, functions are certain mathematical expressions by which it is fitting to articulate states, events, and changes in nature; and mathematical symbols which emanate from contexts of physics are the letters and syllables of the "Language of Science" within which the mathematical functions of physics are not only mathematically coherent but also physically meaningful

---

[6] Thus, H. Weyl in his *Philosophy of Mathematics and Natural Science* (1949), says on p. 8: "Nobody can explain what a *function* is, but. . . ."

I venture to suggest that textbooks and guidebooks for "New Math" in schools, if finding the need for describing functions as correspondences, relations, and suchlike, also ought to take some pains to somehow bring home the fact, in syntax of school young-sters, that no such description is "really" a definition of the notion of function, by which to identify the notion self-sufficiently.

[7] *Decline of the West*, Vol. I, Chap. 2, "The Meaning of Numbers."

and productive. Mathematical functions in physics reveal and control causation, subordination, and other relations; they measure and appraise complexity and simplicity of events; their behavior expresses equilibrium or instability of a state of a system; and they assess generalizations, specializations, and individualizations.

The rise of mathematical functions during and after the Renaissance was intimately connected with the rise of analysis in mathematics. But eventually functions spread into all parts of mathematics and into many areas of rational thinking altogether. The Greeks had never "found" functions in a genuine sense, either in mathematics or in other parts of their system of knowledge. By outward appearance Greek mathematics was geometrical rather than analytical, and by inward structure it was representational rather than operational. Modern accounts of Greek assertions in science, especially of certain statements in the second half of Aristotle's *Physica,* sometimes express them in terms of present-day mathematical formulas. But it may be said that the role of mathematics in physics is not exhausted by merely letting mathematical schemata correspond to so-called physical realities. What is decisive is that suitable mathematical operations within the mathematical schemata shall be representative of meaningful connections and relations within the physical settings to which the mathematical schemata correspond; and in such mathematical operations functions are involved almost always.

To this kind of mathematical activity the Greeks were never able to advance, although some of them, like Democritus and Archimedes, may have come within a pulse beat of doing so. If appraised in retrospect, this incapacity of the Greeks may be said to have been one of their major intellectual failures and one for which no convincing explanation has ever been proposed. The usual rationalizations of the gradual decline of the Greek mathematics of Archimedes and Apollonius from political and socio-economic causes do not convincingly account for the fact that Greek mathematics even when in ascent and at the height of its development did not shape itself into an instrumentality for the creation and promotion of a mechanics and physics, as the mathematics of the 17th and 18th centuries did unceasingly.

In modern times a significant development which accompanied the mathematization of physics was the gradual emergence of the notion of a physical system and also a physical model. The Greeks never arrived at such a notion. As already stated in Chapter 4, the nearest to it in their thinking, especially in the thinking of Aristotle, was a certain notion of a "process-in-nature." Aristotle had various terms for his processes," but it is frequently not easy to compare and delineate these terms as to their proper meaning and scope. In most contexts these terms are unsuited for mathematization or any pre-mathematical formalization, even though the context sometimes suggests comparisons with so-called functions of state or characteristic functions of modern physics, especially thermodynamics.

The Stoics had a recognizable orientation toward making notions of physics specific by a formalization that would make them operationally tangible, but they were seeking to locate such a formalization in a "logic of linguistics" rather than in mathematics.[8] Even the Stoics never developed a trend towards a mathematically controlled physics in general, although Greek astronomy and optics were mathematical, and the mechanical achievements of Archimedes were mathematics at its sternest.

The Schoolmen of the Middle Ages (and perhaps even

[8] Max Pohlenz, "Die Stoa," *Geschichte einer Geistesbewegung,* 2nd edn. I (1959), 22–55; Benson Mates, *Stoic Logic* (1961); Antoinette Virieux-Reymond, *La Logique et l'epistémologie des Stoiciens* (195–).

some late Hellenistical natural philosophers) did have an anticipatory trend toward a mathematization of physics. The Schoolmen were groping for a notion of function, both for mathematics and for physics, and they were trying to inject the notion of mathematical function into a physics which was still largely Aristotelian. Apart from this they were also pursuing a certain extramathematical linguistic "logification" of knowledge after the manner of the Stoics. But one must not exaggerate the sum total of their achievements.

The Renaissance did little for physics but much for mathematics. In mathematics it developed a "school syllabus" which in addition to Euclid's geometry embraced arithmetic, algebra, *Rechenkunst,* combinatorics, and eventually also trigonometry and logarithms. Trigonometry had been a body of knowledge of considerable standing, but it had been a part of astronomy and geodesy. The Renaissance, however, began to study trigonometry as an area of mathematics in its own right, and in a certain vague sense it began to view the addition theorems for sin x and cos x as functional equations per se. Logarithms may have begun as an aid to trigonometry, but they in turn soon showed a tendency to become "independent." Altogether the general study of trigonometry and logarithms, as begun in the Renaissance, was a first phase in the study of the kind of analysis in which functions and manipulations of functions are the central issue. (Present-day reformers of training programs in mathematics ought to consider, and reconsider, that in eliminating trigonometry and logarithms they would abandon an area of mathematics on which an organic understanding of the concept of function has been seeded, grown, and harvested for the last 500 years.) Apart from all this, the Renaissance also edited, printed, and propagated the works of Aristotle, Euclid, Archimedes, Apollonius, Pappus, and others, all of which were studied and exploited in the 17th century as never before and then recast into the style of analysis in due course.

The 17th and 18th centuries created modern physics by creating and giving first preference to a central division of physics which they themselves called Rational Mechanics and which comprised not only mechanics of finite systems and of rigid bodies but also mechanics of continua, whose main divisions are hydrodynamics, acoustics, and elasticity. Physics proper, that is, the theories of light, heat, and electricity and magnetism, were not forgotten or even slighted, but they were advanced at a gentler pace until their turn for full attention came, in the 19th century. Thus Newton composed not only his formidable *Principia* but also an insinuating *Opticks,* at first not in Latin but in English, so that even poets might read it, as some did.[9]

The mechanics of the 17th and 18th centuries is virtually inseparable from mathematics. Beginning with the 18th century the mathematics with which mechanics was most closely intergrown was largely analysis of functions, and virtually all of this analysis was then made for mechanics or because of mechanics. With the beginning of the 19th century, mechanics became a model of mathematization for all of physics and gradually for other science as well. At the same time the analysis of functions began to widen and to pursue aims of its own, but nevertheless throughout the length of the 19th century a considerable part of it continued to be stimulated by physics directly.[10]

---

[9] It is worth noting that the *Opticks* even offers a thermo-radiative description of "black bodies" (the name is in Newton) in which the later definition of G. R. Kirchhoff and Max Planck is anticipated. In fact, "Query 6" in the Appendix to the *Opticks* reads as follows: "Do not black Bodies conceive heat more easily from Light than those of other Colours do, by reason that the Light falling on them is not reflected outwards, but enters the Bodies, and is often reflected and refracted within them, until it be stifled and lost?"

[10] A large amount of information on the interrelations between the growth of analysis and the growth of mechanics and physics is embodied in the great work of H. Burkhart, *Entwicklungen nach oscillirenden Functionen und Integration der Differentialgleichungen der mathematischen Physik,* Jahresbericht der Deutschen Mathematikervereinigung 10_2 (1901–1908), 1800 pp.

6. THE FIRST OCCURRENCE OF FUNCTIONS

According to Moritz Cantor [11] the word "function" was used as a mathematical term for the first time by G. Leibniz in 1694, and even then not immediately in the present-day sense. It may be asserted, however, than an operational conception of function was in the thinking of mathematicians considerably earlier. An eminent geometer of the 20th century, the late G. Castelnuovo, has pointed out [12] that as early as 1604, that is, 90 years before the above given date, the mathematician Luca Valerio, in a study on an Archimedian theme, had *de facto* introduced the notion of a rather general class of (continuous) functions $f(x)$ on a finite closed interval $0 \leqq x \leqq l$. Valerio postulates, in words of his, that the function shall be monotonely decreasing from a positive value at $x = 0$ to the value 0 for $x = l$, and that nothing else shall be assumed. Castelnuovo remarks to this that Archimedes, in accord with other Greek geometers, "considers only curves whose construction or definition is already known (conics, spirals, etc.)." We concur with this judgment of Castelnuovo's, but we must state than an elaboration of this judgment would require a comprehensive analysis. In fact, Archimedes himself gave a celebrated definition of what a "convex curve" is,[13] and this definition, if taken in isolation and by its formal wording, does not require constructibility or any other particularization. But, for instance, the very late

[11] *Geschichte der Mathematik,* III, 2nd edn. (1901), 215–216.

[12] Guido Castelnuovo, *Le origini del calcolo infinitesimale nell'era moderna* (1938), p. 11.

[13] Archimedes, *On the Sphere and Cylinder* I, Postulate 2. The text is as follows. "I apply the term *concave in the same direction* to a line such that, if any two points on it are taken, either all the straight lines connecting the points fall on the same side of the line, or some fall on one and the same side while others fall on the line itself, but none on the other side." Archimedes also gives an "obvious" extension of this definition to surfaces.

commentator Eutocius (A.D. 500) [14] takes it for granted that Archimedes must have been thinking of curves which are put together of pieces of straight lines, circles, conics, etc., that is, of pieces of curves already known.

The *Principia* of Newton had an Archimedean *mise en scène,* and this kept point functions $y = f(x)$ out of sight. In the course of the 18th century, however, mechanics became totally analytical, and by the end of the century point functions, in one and several variables, were as ever-present and all-dominant as they are in all parts of physics today.

In many contexts of physics a function is expected to be given or constructed as an expansion (that is, as an infinite series) of a certain kind. The 17th century immediately introduced expansions in power series, and they originated in mathematics directly but not yet in physics. But power series are the identifying "symbol" of analytic functions, and analytic functions occur everywhere, including physics. The 18th century introduced expansions in trigonometric functions and in related functions, specifically in Bessel functions.[15] These expansions were instigated mainly by requirements of mechanics of continua. They fall under the general notion of expansions in so-called eigenfunctions, and these are the leading instrumentality for the solution of boundary value problems in mathematical physics. If we collect all leading boundary value and eigenvalue problems into one theory, then we may say that a large part of analysis of real variables falls under this theory indirectly, and

[14] His commentary is included in Vol. III of the *Works of Archimedes* by L. Heiberg.

It is a mark of the greatness of a mathematician (and perhaps also of a scientist in general) that he "instinctively" tends to articulate concepts and assertions in formulations which will retain pertinancy in subsequent developments after him, as did Archimedes in the definition which we have quoted in the preceding note.

[15] See C. Truesdell, *The Rational Mechanics of Flexible or Elastic Bodies,* Introduction to *Leonhardi Euleri Opera Omnia,* Ser. 2, Vols. 10, 11 (1960); p. 159 and later passages.

that a large part of the total body of mathematical physics can be viewed to fall under it directly.

It ought to be pointed out, however, that an insight into orthogonality properties of eigenfunctions was slow in coming, and was gained in the 19th century only. This held back the emergence of the Parseval equation which controls the connection between energy and square sum of amplitudes; and without this connection a major part of physics in detail is frustrated. It probably would be idle to speculate whether physics was retarded because orthogonality properties were not available, or whether orthogonality properties were slow in emerging because there was not enough physics to stimulate the emergence.

In the course of the 19th century and afterward, the notion of an eigenfunction became very familiar to physicists; and when, after 1926, E. Schrödinger (and also L. de Broglie) was able to transliterate the then latest version of quantum mechanics from the idiom of matrices into the idiom of eigenfunctions, the comprehension of the new theory and its elaborations were greatly accelerated. This did not mean that the original Heisenberg idiom of matrices was neglected, let alone abandoned. On the contrary, the parallelisms between the two approaches deepened the understanding of each, and a third approach, via operators in Hilbert space, fused the two even more closely.

7. EQUATIONS OF STATE

In the 19th century, most of the growing physics, in electricity and magnetism, in optics, and in heat conduction (as in the work of Fourier), was mathematically modeled on paradigms from mechanisms, and predominantly from mechanics of continua. Therefore, in many parts of physics the mathematics was uniformly the same, not only in technique but also in the manner in which mathematical and physical conceptions were correlated with each other.

However, there was one part of physics which did not conform to this general pattern, and this was the theory of thermodynamics. This became very manifest in the second half of the 19th century when the thermodynamics of E. Clapeyron and R. Clausius metamorphosed into a kinetic theory of matter, which in its turn was mathematically linked to a novel kind of mechanics, namely the so-called statistical mechanics.

Now, this statistically oriented thermodynamics was a counterpart to and descendant of a somewhat earlier thermodynamics, which by no means has been superseded by it. This is the so-called classical or axiomatic thermodynamics, and it has the distinction that, in its "pure" form, it is not directly subordinate to the mechanics of the 18th century, mathematically or even conceptually. It does not obtain its functions by solving differential equations, as does mechanics, but it posits them directly, and these are the "equations of state" or "characteristic equations." Chief among these is the relation

$$pv = RT \tag{5}$$

for a perfect gas which, for $T = t + 273$, was notationally "standardized" by E. Clapeyron in 1834[16] and, for T as absolute temperature, by Lord Kelvin later; but in its content, and by slightly different formulations, the formula had been well known before, inasmuch as it fused the law of Boyle and Mariotte (17th century) with the law of Charles and Gay-Lussac (1802), and also with findings of Amontos (1699). Relation (5) and similar relations do not involve coordinates of space and time, and there is no express reference to the three-dimensionality of the spatial volume v of unit mass in the formulas. In the textbook of Max Planck, *Vorlesungen über Thermodynamik,* no

[16] In his "Sur la puissance motrice de la chaleur," *Journal de l'École polytechnique,* Vol. 14, p. 153.

theory of mechanics is presupposed or developed in passing.

III. THE SIGNIFICANCE OF REAL NUMBERS

8. QUALITY AND QUANTITY

It is generally asserted that modern physics (and science, even social science) is largely quantitative and that older physics was rather more qualitative. This is indeed so, but we wish to say that the distinction between the notions of quantity and quality is subject to some variations of interpretation, and that the determination of the actual meaning of quantitativeness in physics is itself a topic of physics, or at any rate is involved in some major topics of it. Thus, in the theory of relativity, the problems of defining and determining the meaning of temporal simultaneity and of deciding simultaneity by so-called measurements do in a sense involve the over-all problem of the meaning of quantitativeness of physics; and so do, in another sense, in quantum theory, the problems of determining the meaning of observability and of determining the bounds of optimal observation of several magnitudes jointly (uncertainty relation).

We think that the Middle Ages tended to view the terms "quantity" and "quality" as representing contraries of antithetical status, but that in Greek philosophy this was not so, or not yet so. In Aristotle, whenever the notions of quantity and quality appear in the same context, they usually are what Aristotle calls "categories." These two categories, qua categories, are indeed very different from each other, but we think that there is no presumption of contrariety between them. Furthermore, Aristotle usually has a table of categories before him which he sometimes enumerates. The table varies but it never consists of quantity and quality only; even in *Physica,* Book V, Chapters 1 and 2, there is the additional third category of "place" left in the table after others have been discarded.

Finally, in modern physics, reflections on the meaning of quantitativeness are usually directed only toward determining it by itself as such, and comparisons with other "categories" are made only for purposes of explanation and illumination. Last, in working physics, a chain of reasoning or the result of a chain of reasoning is termed "qualitative" if it is intended to be only heuristically suggestive, or "not-yet-quantitative" from whatever cause.

The demand of quantitativeness in physics seems to mean that every *specific* distinction, characterization, or determination of a state of a physical object, and the transmission of specific knowledge and information, must *ultimately* be expressible in terms of real numbers, either single numbers or groupings of numbers, whether such numbers be given "intensively" through the medium of formulas or "extensively" through the medium of tabulations, graphs, or charts. Especially, the value of a basic physical magnitude like mass, charge, or energy is given by stating a real number in addition to the "physical dimension" of the magnitude in question. It has been rightly emphasized by Pierre Duhem that the real numbers which specify such a physical object are usually intended to be the real numbers of arithmetic, which represent the results of some actual or potential "measurments," and not just real numbers which represent the natural ordering of the one-dimensional linear continuum.[17] Thus, if one characterizes the colors of visible light from the limit of red to the limit of violet by real numbers from 7.2×10^{-5} decreasingly to 4.0×10^{-5}, then these are usually quantitative numbers inasmuch as they represent the wavelengths (in centimeters) of the theory of optics. But if one only wishes to represent the successiveness of the colors in the natural linear ordering, then one can suitably label them by real numbers which traverse any finite closed interval

[17] P. Duhem, *La Théorie physique, Son object et sa structure* (1906), p. 179.

in many different ways. But in such a general labeling, the real numbers would only help to establish what medieval philosophers called the "intensity" of color qua "quality" (and not qua quantity). That any quality or condition may be expected to have a certain range of "intensity" of its own was in some sense already known to Aristotle.

The first phase of the rise of quantitativeness in physics was the development of the fully analytical setting for the entire body of rational mechanics in the course of the 18th century; and this development in physics had been preceded by, and then became intergrown with, a development in mathematics itself. The latter development was the arithmetization of geometry, that is, the introduction of coordinate systems into Euclidean space, which, to an extent, had been inaugurated by Descartes, although put into practice considerably later.[18] By these twin developments the notion of physical position in space became fully quantitative, and the vectorial objects of the Newtonian system, namely, velocity, acceleration, momentum, and force, became fully quantitative too. Euclidean coordinates were soon followed by curvilinear coordinates, in mathematics and in mechanics conjointly, and a pioneer architect of them was L. Euler. In his kinematics of rigid bodies he introduced coordinates for the sphere and for the group space of orthogonal transformations, and these coordinates became the prototype of coordinates for certain general group spaces and homogeneous spaces very much later. A notion, or pre-notion, of spaces in more than three dimensions and of general coordinates in them is involved in the Lagrangean mechanics of finite systems with general constraints, and such a general theory could not have been articulated in the pre-quantitative Archimedean setting of the *Principia,* say; although it must be said that Newton, as also Huygens, does reason masterfully within his setting

[18] For the slowness with which the actual arithmetization of geometry proceeded, see the account in C. B. Boyer, *History of Analytic Geometry* (1956).

whenever a constraining condition happens to occur within a context.

In fact, there are quite a number of problems about finite mechanical systems with constraints in the second half of Book I of the *Principia,* and it is breathtaking to contemplate with what unerring self-assurance Newton is able to reason from some made-up geometric figure of Archimedian cast, which for us today is more removed from our mathematical receptivity than is a Homeric battle scene from our poetic sensibility.

Beginning with the 19th century the arithmetization of geometry initiated a systematic development of differential geometry and of topology of general manifolds, and we think that the insights which have been accumulated in these relatively recent doctrines of geometry are in their sum already comparable to those which have been accumulated in the non-arithmetized geometry of the Greeks since Thales. Also, a particular manifestation of the arithmetization of geometry was the establishment of a proper theory of real numbers in the 19th century. It had many features of the theory of proportions for incommensurables in Euclid's famed Book 5. The basic difference in outcome is that while the Greek theory is intended to be a theory for a large class of magnitudes simultaneously, the modern theory constructed one particular magnitude, namely the real number, and envisages a mathematical isomorphism between this particular magnitude and the magnitudes of the large class. Mathematics also has many types of magnitudes, which are not at all isomorphic to real numbers, and in mathematics there is no such compulsion as in physics to represent adequately such magnitudes by groupings of real numbers.

9. VECTORS AND TENSORS

If the various states of a physical system are presented not by single real numbers but by groupings of several real numbers, then usually the elements which constitute a grouping cannot be adequately interpreted singly, each by itself, but must be kept together, either all of them inseparably or at least in certain subgroupings. The various elements of such a subgrouping constitute one indivisible mathematical object, and the elements by themselves are the (real-valued) components of it. The most familiar mathematical objects of this kind are vectors and tensors. They are defined at points of coordinate spaces. The values of the components depend on the particular coordinate system of the space, and there are familiar rules for transformations of components under changes of the coordinate system. Until not long ago it was generally felt that vectors and tensors are, by their essence, mathematical objects of a special kind and for special purposes. But contemporary mathematics tends to subsume the familiar vectors and tensors under a much more comprehensive notion of mathematical objects of several components in general, and this tends to make vectors and tensors into mathematical objects of much less specialized standing.

We have already stated in Chapter 5, that it was the Isaac Newton of the *Principia* who posited that, in his mechanics, velocity, acceleration, momentum, and force should be vectors, thus creating the notion of vector in the process. Newton was the true originator of this mathematical object, even though a textbook conceptualization of it came very much later, perhaps only in the 20th century. We have also stated that, in this sense, tensors of two indices also originated in mechanics, namely in the mechanics of continuous media. The mathematization of the latter theory was carried out in the first half of the 19th century largely by A. Cauchy, and the latter introduced *de facto* the mathematical objects which represent the physical pressure and stress tensor and the accompanying deformation tensor. The name "tensor" for these objects of Cauchy was coined by the crystallographer A. Voigt towards the end of the 19th century and came into general use after 1910.

Formal tensor analysis became generally known when it was chosen as the mathematical idiom of the general theory of relativity, which came into being after 1913. But the foundation of tensor analysis was already laid in the last part of the 19th century (mainly by G. Ricci, but also by T. Levi-Civita, E. Beltrami, and others before them), and its motivation came from physics even then. In fact, on the one hand, the antecedents of the Ricci calculus were in the differential geometry of C. F. Gauss, B. Riemann, and E. B. Christoffel, which had strong links to the Lagrange–Hamilton mechanics of finite system. And, on the other hand, the avowed purpose of Ricci, and his predecessors, in shaping the new calculus was to be able to formulate the partial differential equations of general physics, especially those of elasticity and of electromagnetism, in general curvilinear coordinates of Euclidean space. Thus, the Ricci calculus was intended from the beginning to become a means for expressing mathematical invariance of physical laws, many years before Einstein had adapted it to such a purpose of his own.

Furthermore there was still a certain mathematical development which interposed itself between the work of Ricci and its application by Einstein, and this was the work of Minkowski in 1907–1908 on the adaptation of the theory of Maxwell from Euclidean space-time to "Minkowskian" space-time. Minkowski apparently did not know the work of Ricci, because he starts out from the ordinary calculus of vectors (and tensors) as it was then generally known from potential theory and electrodynamics. Minkowski adapted the notion of ordinary vectors and tensors in Euclidean space to his own four-dimensional space, and he then constructed *in vacuo* a tensor of order 2 which brought together mathematically the Maxwell stress tensor, the Poynting flux vector, and the Kelvin energy density. This was mathematically a decisive step, and although the later theory of Einstein was much more comprehensive, both physically and mathematically, yet a crucial first step had already been taken. Also, the mathematical achievement of Minkowski was very clear cut, and we think that in the theory of relativity only the construction of the relativistic electron by P. Dirac, which came twenty years later, compares with it in incisiveness.

10. IMPORTANCE OF SCALARS

Although physico-mathematical objects of more than one component have become indispensable for expressing the richness of physics and science in all its parts, the scalar objects of one component, that is, basic "magnitudes" of Greek vision, are still the leading entities. First among these is the notion of energy. To Aristotle it was one of the most inscrutable notions in all of natural philosophy. The Rational Mechanics of the 17th and 18th centuries was not equipped to cope with it, even within the confines of its circumscribed aims. Only the physics of the 19th century was able to bring it under cognitive control. But it required all the resources of thermodynamics to accomplish this, and much of the resources of electricity and magnetism too. In the 19th century, energy became one of the most talked-about concepts in all science and also in philosophy. Industrial and social revolutions were waged in its name, and Marxist analysts take pride in the (alleged) fact that Friedrich Engels was an important champion of it.

Finally we wish to point out that one of the most radical innovations in physics was the outcome of a problem regarding energy which had been quietly building up throughout all of the 19th century. The problem was the problem of radiation (which is a certain mode of transportation) of energy across "empty" space, and this problem cut across several divisions of physics, thermodynamics, optics, and finally also electromagnetism (Poynting flux vector); and the preoccupation with this problem led at the very end of the 19th century to the creation of quantum theory, to which the future of physics belongs.

Temperature and entropy are a unique pair of scalars, which seem to belong together mathematically. Temperature is at first consideration an everyday "observable," and yet physics required 200 years to conceptualize it. Entropy is not an outward "observable" but on the contrary an "internal" construction, and yet it is as controversial today as it was on the day it was conceived.

Even the concept of mass, which is the scalar of all scalars, was firmly set down only by Newton in his *Principia,* after irritating natural philosophy for over 2,000 years. And the concept of mass continues to be very mystifying. Thus the assertion in physics that mechanical mass and gravitational mass are equal will never cease to baffle the "layman," and the mystique of the statement that mass and energy are one and the same, $E = mc^2$, will never cease to stir his imagination, not to mention the fact that professional physicists themselves would feel easier if they could tell how the "petrification" of energy into clumps of masses of mysteriously foreordained elementary-particle-sizes actually comes about.

Finally we venture to suggest that Greek mathematics was both made and unmade by the efforts of the Greeks to conceptualize simple scalar magnitudes like length, area, and volume. The Greeks created the *eidos* of mathematics, and it is this which made the *Elements* of Euclid one of the greatest best sellers of all times, in spite of the tediousness of its schoolmasterly style. But *creatively* the Greeks were more philosophers and poets than mathematicians, and they did not have in their creativeness the instinctive realization that in the actual creation of mathematical subject matter the complex comes before the simple, the recondite before the obvious, the hidden before the manifest. Greek mathematics discovered almost at birth some deep-seated problems of the *eidos* of mathematics, one of which was the problem of the true mathematical structure of scalar "magnitudes." Modern mathematics postponed this problem to the 19th century and was none the worse for it. But Greek mathematics compelled itself to face the problem fully from the beginning and exhausted the strength of its youth in mastering it prematurely. Archimedes could, within his setting, define the concept of a "limit" as nobody in modern mathematics before Cauchy; and Newton was totally unable to define rigorously instantaneous velocity and instantaneous acceleration to the end of his days. And yet Newton created analysis and mechanics, and Archimedes did not; and Newton was the first mathematical physicist, and Archimedes was not. We note that the honor of having been the first mathematical physicist ought perhaps to be divided between Newton and Huygens, if not even bestowed on Huygens exclusively. Huygens was indeed a physicist among the greatest, and in his work on evolutes in his "Horologium" he virtually shamed Apollonius of Perga by using Apollonius' own syllogisms, and *seemingly* no others, toward conceiving of a "local" curvature for a general convex curve; whereas Apollonius himself, in the fifth book of his *Conics,* came near the concept of curvature and went repeatedly around it but did not locate it. But it is a fact that Newton did compose a *Principia* for the ages, and Huygens did not, just as Charles Darwin did compose an *Origin of Species,* and H. R. Wallace did not; and such facts do count.

IV. THE SIGNIFICANCE OF COMPLEX NUMBERS

11. THE COMPLEXIFICATION OF PHYSICS

We think that it is legitimate to say that in a certain broad sense, the problem of the significance of the continuum of real numbers for physics has been a perennial major topic of Western philosophy (including medieval Arabic philosophy) from around 500 B.C. until the present. In Greek philosophy, aspects of this problem reflected themselves in views of Pythagorean circles that "all is number"; in the renowned and (to philosophers) timeless paradoxes of Zeno of Elea; in the Platonic doctrine of great-and-small (perhaps); and in the multiform analyses of the meaning of Infinite, Space, Time, and Motion in Aristotle.[19] Through the permanency of Greek topics in later philosophy, and through other impulses, the problem of real numbers continued to recur in all stages of natural philosophy afterward.

However, in addition to real numbers contemporary physics also employs complex numbers. These numbers have been known to physics for 200 years, and in the 19th century their occurrence in physics has gradually widened to such an extent that a veritable "complexification" of physics has come about. However, there seems to be no recognized topic or subtopic in philosophy whose subject matter is to evaluate the significance of complex numbers in mathematics, physics, and science in general. Even the far-ranging philosophies of mathematics of Bertrand Russell and A. N. Whitehead in the first half of the 20th century were apparently not concerned with this situation.[20] We do not intend to undertake such an evaluation ourselves, but we venture to make some observations on the manner in which the "complexification" of physics expresses itself.

Complex numbers are pairs of real numbers, and by taking real and imaginary parts, any set of relations in mathematics and physics which involves complex numbers can be replaced by an equivalent set in which only real numbers occur. However, as in the case of real numbers, so also in the case of complex numbers what is crucially decisive for physics is the fact that they can be multiplied

[19] Most of what is known about Greek views on the nature of mathematics comes from the *Physica* and *Metaphysica* of Aristotle.

[20] See, however, a recent article by E. P. Wigner, "The Unreasonable Effectiveness of Mathematics in the Natural Sciences," *Communications on Pure and Applied Mathematics,* Vol. 13 (1960) pp. 1–14.

with each other. We saw at the beginning of this chapter how the 17th-century mechanics created the all-important conception of a momentum by doing, mathematically, no more than multiplying any two real numbers with each other. In the case of complex numbers the corresponding situation is even more remarkable and philosophically unfathomable. We can do no better than simply report it as a fact that the manner in which the real and imaginary parts of complex numbers are linked with each other by the rule of multiplication somehow endows the system of complex numbers with a range of power, effectiveness, and suppleness which in mathematics itself is both inexhaustible and inexplicable, and which has been affecting the course and development of physics deeply and even radically. It is true that a contemporary physicist usually accepts complex numbers as something comfortably familiar and marvels only at mathematical operators and groups, if at all. But we wish to say that the operators which occur in physics are virtually all intergrown with complex numbers, and that groups which occur in physics, if they are Lie groups, are also linked with complex numbers, in their mathematical background at any rate.

In the physics of the 19th century, and also in the theory of relativity, which by the texture and spirit of its mathematization was a terminal phase of that physics, complex numbers were being used ever more frequently, both in calculations and in derivations, and they were beginning to penetrate into thought patterns also. But basic conceptualizations and basic formulations continued to be presented and expressed in real variables only. However, quantum mechanics changed even this. The very basic equations

$$pq - qp = \frac{h}{2\pi i}$$

$$\frac{h}{2\pi i} \frac{\partial \psi}{\partial t} = -H\psi$$

of the new theory display the symbol $i = \sqrt{-1}$, which is indeed only a symbol, openly and directly. Also in these and other formulas of quantum theory the mathematical imagery is quite intense, and it suggests more credibly than in previous theories the possibility of attaining penetration into physics through symbolization in mathematics and of finding reality of nature through irreality of mathematics.

By the separation into real and imaginary parts, one can of course formally eliminate complex values from the above equations and also from the computation of the Born probability density out of the wave function. And one can certainly eliminate the imaginary from the de Broglie wave

$$\exp\left\{\frac{2\pi}{hi}(Et - p_x x - p_y y - p_z z)\right\}$$

by taking its real or imaginary part, say. But any attempt to do this in general, and not for a specific purpose on a special occasion, would run counter to a development whose stream is irresistible and whose course is indeflectable.

12. COMPLEX NUMBERS IN THE 18TH CENTURY

The analytical strength of complex numbers began to be recognized towards the end of the 17th century, and the principal analytical findings up to the middle of the 18th century culminated in the "magic" formula

$$e^{ix} = \cos x + i \sin x$$

which was clearly stated by Euler in his "Introductio in analysin infinitorum" in 1748. In this formula the letter x denotes arc length so that in particular

$$e^{i\pi} = -1, \quad e^{2i\pi} = 1$$

[237]

and the great strength of these relations stems from the fact that the three symbols

$$e, \ i, \ \pi$$

which they bring together, are, by *provenance,* unrelated to each other. Such linkages of three mathematical objects of disparate provenance in one formula are exceptional; in general, even centrally productive formulas link up only two separate mathematical objects at a time.

In the second half of the 18th century complex numbers took a short but notable step on a path which, after a considerable halt, began to lead them to an area in which mathematics and physics were frequently to meet in the course of the 19th century. The step was made by d'Alembert.[21] In a work on hydrodynamics he introduced in 1752, and even earlier in 1746, the system of equations

$$\frac{\partial p}{\partial x} = \frac{\partial q}{\partial y}, \ \frac{\partial p}{\partial y} = - \frac{\partial q}{\partial x}$$

which nowadays is named after A. Cauchy and B. Riemann of the 19th century. And d'Alembert stated, in his style, that a general solution of the system can be represented as

$$p = \phi \left(x + iy \right) + \phi \left(x - iy \right) + i\psi \left(x + iy \right) - i\psi \left(x - iy \right)$$
$$q = - \ i\phi \left(x + iy \right) + i\phi \left(x - iy \right) + \psi \left(x + iy \right) + \psi \left(x - iy \right)$$

where, in our parlance, $\phi \left(z \right)$ and $\psi \left(z \right)$ are analytic functions of the complex variable $z = x + iy$ and have real values on the x-axis. In 1755 Euler hailed the result of

[21] See C. Truesdell, *Rational Fluid Mechanics, 1687–1765,* pp. LIV and LV, in *Leonhardi Euleri Opera Omnia,* Ser. 2, Vol. 12, 1954; also Paul Stäckel, *Integration durch imaginäres Gebiet,* Bibliotheca Mathematica Series III, Vol. 1 (1900), pp. 109–128, and *Beiträge zur Geschichte der Funktionentheorie im achtzehnten Jahrhundert,* the same series, Vol. 2 (1901), pp. 111–121.

d'Alembert and observed that the above system of equations is equivalent to the one equation

$$p + iq = 2\phi(x + iy) + 2i\psi(x - iy) \qquad (6)$$

and he gave proofs for all statements involved. We note that the assertion of Euler is valid, according to present-day standards, in the following version. Let $f(z)$ be defined and analytic in a domain D of the z-plane, and let the domain be symmetric with regard to the x-axis. Being connected, the domain D must then also include points of the x-axis. Due to the symmetry of D, there exists a function $\overline{f(\bar{z})}$ in all of D, and it is also analytic, so are therefore the functions

$$\phi(z) = \frac{1}{4}\{f(z) + \overline{f(\bar{z})}\}, \qquad \psi(z) = \frac{1}{4i}\{f(z) - \overline{f(\bar{z})}\}$$

Now, $\phi(z)$ and $\psi(z)$ are real on the x-axis, and also

$$f(z) = \phi(z) + i\psi(z)$$

which is formula (6).[22]

In the 19th century, analytic functions of a complex variable spread into many areas of physics, and the twin topic of two-dimensional potential theory and conformal mapping received all its impulses in this way. But during the first 100 years after d'Alembert's inspired result the tempo of these developments was quite slow.

Another major development which began in the 18th century and also led to complexification of physics in the 19th century was the emergence of trigonometric expansions and related expansions in rising theories of vibrations. During the 18th century the basic functions in trigonometric expansions were the real functions cos nx, sin nx, and not the imaginary exponentials e^{imx}, and during the whole 19th century there were physicists who almost

[22] This very simple proof of the assertion is due to Professor R. C. Gunning; my own proof was complicated.

always used the real functions. However, the complex exponentials were known, and whenever they were demanded imperatively they were used. In the 18th century Euler stated the general theorem, which is rightly named after him, that a solution of a differential equation

$$\sum_{m=0}^{n} a_m \frac{d^m f}{dx^m} = 0 \tag{7}$$

with constant coefficients is a linear combination with constant coefficients of monomials

$$x^\mu e^{\lambda x}$$

where λ is any root, real or complex, of the polynomial

$$\sum_{0}^{n} a_m \lambda^m$$

and where $\mu = 0, 1, \ldots, r - 1$, if r is the multiplicity of the root. Now Euler was led to the investigation of general equations (7) by generalizing the "eigenvalue" equations

$$\frac{d^2 f}{dx^2} = af, \frac{d^4 f}{dx^4} = af$$

to which Daniel Bernoulli had been led in his work on vibrations in 1735.[23]

13. COMPLEX NUMBERS IN OPTICS

When solving differential equations which describe undulatory phenomena by the method of eigenfunctions, it is nearly always mathematically very suggestive to set up the mathematical solutions of the equations as complex-valued functions and then to take, as actual solutions of the physical problems, the real parts of the resulting functions only. This has become such an indispensable routine that it

[23] See reference in n. 15.

has almost developed a habit of thinking in terms of complex-valued physical objects directly. However, in the resulting complexification of the physics of undulatory phenomena, concurrent with this habituation from mere procedural appropriateness, there has also been a motivation from a genuine cause, and the originator of this motivation was A. Fresnel in his theory of total reflection (1823, published 1831).[24]

For propagation of light in adjoining transparent bodies [25] Fresnel had established that if α and β are angles of incidence and refraction, with

$$\sin \alpha = \mu \sin \beta \qquad (8)$$

then for light polarized in the plane of reflection the ratio of amplitudes of reflected and incident light

$$- \frac{\sin (\alpha - \beta)}{\sin (\alpha + \beta)} \qquad (9)$$

If however $\sin \alpha > \mu$, then the angle β as computed from (8) is complex valued. The ratio (9) is also complex valued, but it has absolute value 1, and thus it "represents" a pure shift $e^{i\theta}$, where

$$\tan \frac{1}{2} \theta = \frac{\sqrt{\sin^2 \alpha - \mu^2}}{\mu \cos \alpha} \qquad (10)$$

Fresnel now interpreted this to mean,

sinon par des raisonnements rigoureux, aux moins par les inductions les plus rationelles et les plus probables,

that there is total reflection, that the amplitudes of the two waves are the same, but that the waves differ in phase by the amount θ as given by (10). Fresnel also made a

[24] *Oeuvres d'Augustin Fresnel,* Vol. I (1866), pp. 767–799, especially p. 782.
[25] Compare E. Whittaker, *A History of the Theories of Aether and Electricity,* I (1951), 126.

corresponding statement for light reflected at right angles, and both statements were completely confirmed by experiment.

We think that this was the first time that complex numbers or any other mathematical objects which are "nothing-but-symbols" were put into the center of an interpretative context of "reality," and it is an extraordinary fact that this interpretation, although the first of its kind, stood up so well to verification by experiment and to the later "Maxwellization" of the entire theory. In very loose terms one can say that this was the first time in which "nature" was abstracted from "pure" mathematics, that is, from a mathematics which had not been previously abstracted from nature itself.

Almost until the end of the 19th century efforts were made by some physicists to rationalize the conclusion of Fresnel more "convincingly," and even Henri Poincaré took part in these endeavors.[26] But others were more interested in exploiting the idea of Fresnel for more elaborate problems, that is, for problems in which at least one of the two bodies is optically opaque. From 1837 on, J. MacCullagh and A. Cauchy turned their attention to the then current topic of metallic reflection,[27] and they offered a first theory of the phenomenon by mathematically introducing into formulas of Fresnel a refractive index which is a general complex number $\mu = \nu - i\kappa$. This assumption has maintained itself in later developments, and the real part ν is a "real" refractive index and the imaginary part κ is an index of absorption.

Sometimes even a working physicist pauses to wonder at the extent to which the symbolization by complex numbers is penetrating into the mathematical representation of schemata of physics at their depth. For instance, Arnold

[26] See his *Théorie mathematique de la lumière,* Vol. I (1889), p. 336.

[27] Whittaker (see n. 25), pp. 161–169.

Sommerfeld (1868–1951) was wont to operate freely, from the beginning of his career, with complex-valued solutions of divers wave equations, but apparently in the expectation that in the end the real and imaginary parts of the solutions would have a physical interpretation separately. However, when analyzing a Schroedinger equation for the Zeeman effect of the hydrogen atom, he found that a certain very "natural" solution, which fitted well the case of the so-called normal (that is non-relativistic) Zeeman effect, was such that its real and imaginary parts did not have a physical standing by themselves. He apparently attached importance to this fact. He stated it expressly in 1929 on page 141 of his *Wellenmechanischer Ergänzungsband,* which was a supplement to his leading work, *Atombau und Spektrallinien;* and he fully restated this fact in 1939, on page 104 of Volume 2 of the fifth edition of the entire work, although in the intervening ten years the "complexification" of physics had been progressing even faster than at any time before.

V. THE SIGNIFICANCE OF MATHEMATICAL SPACE

14. THE SPACE-TIME PROBLEM

At present, mathematics is growing and spreading very fast, the so-called applied mathematics is steadily increasing in volume, and in certain areas of physics some difficult modern mathematics is being tried out. And yet one can say that during the last few decades no stirring mathematical innovations have been injected into physics, or have even been provoked by physics, and that, for instance, the far-reaching innovations in nuclear physics have not been reflected in depth in the simultaneous innovations in mathematics which have been equally far-reaching. Apparently the "predestined harmony" between mathematics and physics is itself a phenomenon of variable intensity, and if this is so, it would be most important that philosophy of science should study its variability and determine whether it has "hidden periodicities" and what they are.

We think that the philosophical problem of the relation between mathematics and physics is not receiving sufficient attention. This is due in part to the fact that this problem is a very difficult one, but it is also due in part to the fact that present-day philosophy of nature is still heeding the task which the 17th century has imposed on it.

The science of the 17th century gave a primacy to the theme of space and time which it inherited from late medieval natural philosophy, Western and Arabic. The Kantian philosophical system of the 18th century, and some philosophies preceding it, made this primacy within physics into a paramountcy within all of metaphysics, and this paramountcy has remained unchallenged. The 19th century acquiesced in it, and the theories of relativity and gravitation in the first decade of the 20th century even enhanced it. But quantum theory, which is a profoundly new method in scientific thinking, may eventually change this. It is true that in quantum theory the new vistas on space have been startling and are as yet not fathomable, but the theory nevertheless bids fair to lower the notions of space and time from their Kant–Einstein heights of dominancy to the general level of other notions of the evolving physics of particles, some of whose concepts and phenomena are equally challenging and sometimes inscrutable. It is true that there is a challenging unidirectedness of the "arrow of time" in biology, cosmology, and even thermodynamics, and that this suggests problems which invite commanding attention, but it can be said that in many philosophical works which assert and elaborate on the supremacy of space and time, such concrete problems have not even been featured. Furthermore, in large parts of physics and other science, only mathematical space and time are operationally of consequence, and we think that in this respect the notions of space and time have not been causing undue difficulties. On the contrary, the mathematization of space and time has been usually more advanced than that of other notions and phenomena of physics in the same epoch.

15. SPACE BEFORE ARISTOTLE

According to various remarks in Aristotle, a mathematization of the everyday statement that space is three-dimensional was consciously begun by Pythagoreans, and Plato refers to this in the *Timaeus*. Aristotle tries, in *Categories,* 5 a 8–14, to give a "logically" worded definition that the place (*topos*) of an object is its "extension," and we think that a three-dimensionality of this extension is there implied. Also, large parts of Aristotle's *Physica,* and of other works, deal in one form or another with the problem of mathematization of the statement that space of physics is "continuous," a problem which had apparently monopolized the attention of leading Eleatics. But Aristotle does not make it clear whether Pythagoreans were already concerned with this problem, except that in *Metaphysica* 1083 b 7–19 he reproves Pythagoreans for holding the view that bodies are composed of numbers. Such a view, he says, or they said before him, leads to the conclusion that there are spatial magnitudes which are indivisible, and this is highly repugnant to Aristotle. (In contemporary physics, many an elementary particle has a finite non-zero radius of "influence" associated with it, and is "indivisible" until and unless it is transmuted into or collides with another particle.)

As we have already stated in Chapter 4, it is important to realize that in the thinking of Plato and Aristotle the notions of space and time in their role of "scientific" concepts were objects of physics, and of physics only, and not at all of metaphysics; except that the natural philosophy of Plato was still in the "archaic" state in which physics proper was not separable from cosmology and cosmological genesis. Thus, Plato deals with space and time mainly in the *Timaeus,* and hardly in the ontological dialogues *Theaetetus* and *Sophistes.* And Aristotle deals with them fully in the *Physica* and *De Caelo,* and occasionally in "descriptive" science, but not in the *Metaphysica,* although the latter re-discusses other notions from the *Physica* on a "higher" level of physical primacy, and fully discusses the nature of mathematics and its relations to physics and to other areas of knowledge.

16. SPACE IN ARISTOTLE

We have already commented upon Aristotle's notion, or notions, of space in Chapter 4, especially in section 15 of it. We are now going to make a further observation, and this one will take the present-day role of *mathematical* space into account.

The space of Newton's mechanics may have been a nearly metaphysical three-dimensional *absolute* datum, and the space of Kant was most fully metaphysically so. But it is a fact that the space of Euler, Lagrange, Jacobi, Hamilton, and other architects of the vast doctrine of mechanics in the 18th and 19th centuries was, in its operational aspects, that is, at least during their "working hours," entirely physical and mathematical only, and that, in a certain novel aspect or version, it additionally became something entirely different from what it is to psychology and philosophy even today. Namely, in addition to the three-dimensional ordinary space of common human awareness, Rational Mechanics also gradually introduced the concept of a purely mathematical space which is multi-dimensional, without any restriction on the dimension, which has a general differential geometric structure, and which serves as a substratum of mathematization for a general physical set-up, not only in mechanics but also in physics proper, and in fact in science anywhere.

Such a space is variously called the "space of generalized

coordinates," or "space of parameters," or sometimes "phase space," and it has as many mathematical dimensions as there are physical "degrees of freedom" in the given set-up. The admissible generality of the differential geometric structure of such a space became greatly emphasized in the theory of relativity, but it had actually played a role before, as we have already stated above in section 9. Also, some "models" of infinitely dimensional spaces had already been introduced in mechanics of continua before infinitely dimensional space became the "base space" of quantum theory, and thus an everyday occurrence in it.

Now, we venture to suggest that an anticipation of such a conception of space, albeit only a very dim one, may be imputed to Aristotle in his "essay" on *topos* in *Physica*, Book IV; whereas in Plato only the very transparent and obvious conception of space *à la* Euclid is discernible. And we think that it is Aristotle's half-awareness of this which makes him assert, and with some emphasis, that although Plato had made some correct and satisfactory statements about space, there is still considerably more that can be said about it. At any rate we think that if one does impute such an anticipation to Aristotle, then some additional understanding of Aristotle's approach to the problem of space may accrue. In fact, in the light of such an anticipation, Aristotle's statement that for a circumscribed system the *topos* is "the innermost boundary of the enclosure" would be his way of saying that any state of such a system is determined by a subspace in a comprehensive general space of parameters and that, if the subspace is a domain, an appropriate way of determining the domain is to exhibit its boundary. We note that Greek mathematics did not find it easy to separate terminologically, and to an extent even apperceptively, a domain from its boundary and was inclined to view the boundary as the prime attribute of a domain. Also, in Aristotle's famous case of a boat floating on a river (212 b 15–20), which gave rise to controversy, if one wishes to understand fully the varying physical effect which the water flowing past the boat has on its hull, it does not suffice to consider only a parameter space for the boat per se in isolation; a parameter space for the system consisting of boat and river must be introduced. And this then leads to the variant characterization that *topos* is "the innermost *motionless* boundary of what contains."

17. THREE-DIMENSIONALITY OF SPACE

After Plato and Aristotle, a certain measure of awareness that a mathematization of space is a prerequisite for "exact" science may have been in the minds of some natural philosophers from Aristotle to Galileo, but only the 17th century began to act in the matter.

We will now make observations on the extent to which the three-dimensionality of space, already known to Plato and Aristotle as a scientific fact, manifests itself in mathematical laws of physics. There are general basic laws in which it does not manifest itself; such are, in classical mechanics, the Euler–Lagrange equations when stated in Lagrangian free parameters, and, in quantum mechanics, the general equations of Schrödinger and Heisenberg.

The three-dimensionality of space is involved in Kepler's statement in his *Epitome* (1618–1621) that light is attenuated in the inverse square of the distance from its source.[28] More incisively, but considerably later,[29] Immanuel Kant stated that in Newton's law of gravitation the inverse distance is raised to the power 2 because the dimension of space is 3. In the same context Kant seems to find a reason why the dimension ought to be an odd number; here I wish to recall that in classical physics a

[28] *Great Books of the Western World* (Chicago, Encyclopedia Britannica), XVI, 903.
[29] *Gedanken von der Wahren Schätzung der lebendigen Kräfte* (1747), §10, 11.

preference for odd dimensions over even dimensions occurs in the so-called Huygens principle, which states that in odd-dimensional spaces there are sharp conical light signals as solutions of the wave equation, whereas in even-dimensional spaces the signals are cylindrically diffuse.[30]

A significant characterization of three-dimensionality of space is embodied in Book I of Newton's *Principia*. After (very important) preliminaries, Newton derives in "Proposition I. Theorem I." of Section II [31] the general law that in any field of central forces (with one center) a mass point describes an orbit which lies in a two-dimensional plane and that the area which the radius vector from the center to the mass point sweeps out in this plane is proportional to the time elapsed. Now, Newton reasons in an Archimedean setting in such a way that his reasoning applies to a central force field in a Euclidean space R^n, for any dimension n; therefore, Newton's derivation of the three planetary laws applies to the motion of a mass point in a force field $1/r^2$ in R^n, for any $n \geq 2$. However, in Section XII ("The attractive forces of spherical bodies") of Book I[32] the three-dimensionality of space becomes indispensable; the representative context in which it enters is the assertion that a sphere of finite radius whose density is a function of the distance from the center of the sphere acts, gravitationally, outside itself as if its entire mass were concentrated in its center. For the force $1/r^2$ this assertion holds in R^3, and in no other [33] R^n.

In classical physics in general, the three-dimensionality of space manifests itself in a formulaic pecularity. If a two-

[30] R. Courant and D. Hilbert, *Equations of Mathematical Physics*, Vol. II (1962); see index under "Huygens Principle."

[31] See p. 40 in the edition of the *Principia* by F. Cajori (University of California Press, 1934, 1962).

[32] Beginning on p. 193 of the last named edition.

[33] A. S. Ramsey, *An Introduction to the Theory of Newtonian Attraction* (1940), p. 62, problem 49.

index tensor H_{ij} is skew symmetric, that is $H_{ij} = -H_{ji}$, then in R^n it has effectively $n(n-1)/2$ components; in R^3 the latter number is 3, so that, in R^3, a skew-tensor is a vector, but a conceptually different one. In electrodynamics the laws of induction implied the following distinction between the electric vector E_i and the magnetic vector H_i; if one of them is translational then the other is rotational. Kelvin and Maxwell adopted the premise that H_i is rotational,[34] and this premise has never been challenged; however the actual "operational" identification of the magnetic force H_i with a skew-tensor H_{ij} was slow in coming. It was formalized in Minkowski's great memoir of 1908,[35] but even then it began to be generally known only after full-length books on the theory of relativity had begun to appear. Finally, Lorentz–Minkowski space underlies the present-day quantum field theory, and its $(3+1)$-dimensionality manifests itself in the particularity of various group representations occurring.

In the 19th century, the fact that the space of cognition is three-dimensional was a recurrent problem of the then Theory of Knowledge, although a somewhat marginal one. More significantly, the visual arts carried out a program of de-emphasizing the three-dimensionality (or, in fact, any dimensionality) of space, which culminated in the achievements of Paul Cézanne;[36] but these efforts in the arts have apparently not, or not yet been brought into parallel with generalizations of the conception of space in Lagrangian mechanics, set theory, topology, etc.

18. MATHEMATICAL SCHEMA OF TIME

Mathematically, time has rarely[37] been envisaged to be anything other than a simple one-dimensional interval,

---

[34] See Whittaker (n. 25), p. 247.

[35] Whittaker, Vol. II, p. 64ff.

[36] Erle Loran, *Cézanne's Composition*, 2nd edn. (1944).

[37] For the only major exception see G. J. Whitrow, *The Natural Philosophy of Time* (1961), pp. 256–261.

either a finite interval (a, b), or the half-line $(0, \infty)$, or the entire line $(-\infty, \infty)$; except that in Lorentz–Minkowski $(3 + 1)$-space the time and space components are inseparable under Lorentz transformations. However, there is the following complication to which there is no analogue in the space problem. Time "elapses," and the only general way of "fixing" it is through mensural determination by means of a phenomenon which is presumed to be cyclical, that is, "uniformly" periodical. Such a periodic phenomenon has, in addition to the universal time whose schema is a linearly strictly ordered interval, an individual time whose schema is a circle. And, as seen in retrospect, there is already in Aristotle a groping for the insight that to measure time by a periodic event means, in terms of topology, to perform a locally one-one continuous mapping of the universal time interval into the individual circle. The Greeks did not have a terrestial pendulum, let alone an atomic or even biological rythmic pulse. But they had the revolution of the heavens, and Plato knowingly identified time with this revolution. In this he was anticipated by the Pythagoreans, who, in a gauche manner, identified time with the heavenly sphere itself; this conceptual inadvertency brought down upon them the sarcasm of Aristotle, quite unjustly. Aristotle also toyed with the possibility that there might be, simultaneously, separate time scales in separate universes, that is, in our parlance, in separate clusters of galaxies; but he dismisses this possibility peremptorily, as he sometimes does with his uncanny anticipations.[38]

Some medieval philosophers strove to harmonize the Aristotelian doctrine that the universe is ungenerated, meaning that the universal time fills the entire axis $(-\infty, \infty)$, with the Biblical doctrine that there was a genesis, meaning that the universal time is the half-axis $(0, \infty)$; and

---

[38] To the above see Aristotle's *Physica*, Book IV, Chapter 10, with commentaries in the edition of W. D. Ross, and, importantly, in the Loeb Edition.

they tried to bring about harmony through the interesting suggestion that significant world events in the half-line (0, ∞) have some kind of pre-existence, potentially, in the anterior half-line (−∞, 0). Mathematically this raises the question whether significant world functions on the half-axis of time (0, ∞) have a unique continuation backward into the half-line (−∞, 0). It is interesting to note that such a continuation is indeed possible for general (non-analytic) periodic, and more general almost periodic, functions, and for large classes of solutions of ordinary and partial differential equations of physics, especially for solutions of the d'Alembert–Maxwell wave equation. But this cannot be so asserted for solutions of differential equations of the heat equation type, or, apparently, for functions which describe the behavior of particles in contemporary nuclear theories of matter. It is this, above all, which makes it physically—and perhaps also meta-physically—meaningful to assert that our universe has come into being a *finite* time ago, by some spontaneity ("big bang" theory).

19. THE FUTURE OF THE SPACE-TIME PROBLEM

Since the 17th century the intellectual orientation has become such that problems about space and time command overriding attention, as a matter of course. And yet we think that for the comprehension of the universe it is just as important to know how the reality of physics can be articulated by the symbolism of mathematics as it is to know, beyond the shadow of a doubt, what space and time are. The Greeks had little mathematics and less physics and no symbols, and yet Plato and Aristotle gave much less weight to the problem of space and time than to the problem of the relation between mathematics and physics or, what was virtually the same to them, than to the problem of the relation between mathematics and knowledge in general. From Plato onward, Greek *mathesis* was "mathematics" and "general knowledge" both. Leibniz tried to activate this latent ambivalence of the meaning of *mathesis,* but, although he received a respectful hearing, not much ensued. The second half of the 20th century is about to try again, and this time there seems to be a gathering mood for a comprehensive accord on the need of a *mathesis* of universality at the root.

CHAPTER 7

THE ESSENCE OF MATHEMATICS *

1. RELATION TO SCIENCE

Mathematics is frequently encountered in association and interaction with Astronomy, Physics, and other branches of Natural Science, and it also has deep-rooted affinities to what are called Humanities nowadays. Actually, it is a realm of knowledge entirely by itself, and one of considerable scope too; the word "mathematics" stems from a root which means Learnable Knowledge as such. Mathematical knowledge is commonly deemed to have a high degree of validity, binding on *Homo sapiens* irrespective of cultural conditioning and predilection, although it can be argued that in the past cultural settings have affected its development noticeably.

Even as far as the scientist is concerned, mathematics is not a branch of natural science itself. It does not deal with phenomena and objects of the external world and their relations to each other, but, strictly speaking, only with objects and relations of its own imagery. Mathematical figures in two- or three-dimensional geometry are largely idealizations of objects occurring in the physical world, but figures in n-dimensional space for general n no longer are. Integer numbers 1, 2, 3, . . . and even real numbers in general can be claimed to be abstractions from quantities occurring in the physical world, but the "imaginary" number $i = \sqrt{-1}$ has received its name, still current, from the very fact that it no longer so is, even though the use of complex numbers $a + bi$ is indispensable to science nowadays. It a laboratory experiment idealizes a physical

* Reprinted, with adjustments, from the *McGraw-Hill Encyclopedia of Science and Technology,* Copyright © 1960 by McGraw-Hill Inc. Originally entitled "Mathematics."

system in order to eliminate secondary features not essential to the study at hand, its ultimate objective is an understanding of the un-idealized physical system nevertheless. But even if it were true that the five regular solids as investigated in the 13th book of Euclid's *Elements* (cube, tetrahedron, octahedron, icosahedron, dodecahedron) had been found due to the occurrence of approximating crystals in nature, nevertheless, once found, the idealized geometric figures become primary objects of mathematics which are definitive as such.

Mathematics is not subordinate to natural science by being a handmaiden of it, and one can practice competently meaningful mathematics without being concerned with science at all. Especially, philosophical attempts to reduce all origin of mathematics to utilitarian motives are wholly unconvincing. But it is fair to say that mathematics is the language of science in a deep sense. Mathematics is an indispensable medium by which and within which science expresses, formulates, continues, and communicates itself. And just as language of true literacy not only specifies and expresses thoughts and processes of thinking but also creates them in turn, so does mathematics not only specify, clarify and make rigorously workable concepts and laws of science which perhaps, partially at least, could be put forward without it; but at certain crucial instances it is an indispensable constituent of their creation and emergence as well. In Newton's formula for the motion of a particle on a straight line

$$m \frac{d^2x}{dt^2} = F$$

the mass m and the force F are non-mathematical objects perhaps. But the instantaneous velocity $v = dx/dt$ and the instantaneous acceleration $a = dv/dt = d^2x/dt^2$ are wholly mathematical, and without a mathematical theory of the infinitesimal calculus not conceivable. Newton the physicist was driven to creating his version of the calculus because of this. Also, Newton had to have not only the process of differentiation but also the concept of a mathematical function, because only a function can be differentiated. He required not only the path function $x = x(t)$ but, for the second derivative, he had to envisage the velocity $v = dx/dt$ itself again as a function depending on t, even though in the definition of v this dependence had been reduced to "instantaneousness." This concept of function was given to Newton by the then new theory of analytic geometry of Descartes, and it is a fact that after Archimedes' work on the motion of a lever, Theoretical Mechanics stood virtually still for almost 2,000 years until the twin mathematical concepts of function and derivative were ready to emerge.

2. CREATIVE FORMULAS

A formula is a string of mathematical symbols subject only to certain general rules of composition. To a working mathematician a "string of symbols" is a formula if it is something worth remembering. Much mathematics is concentrated in, and propelled by, certain formulas of unusual import.

The oldest such is the Pythagorean theorem $c^2 = a^2 + b^2$, where a, b are the sides and c is the hypotenuse in a right-angled triangle. If a, b are rational numbers p/q, then c need not be so, for instance, for $a = b = 1$, $c = \sqrt{2}$ is not. This compelled the Greeks to study ratios of incommensurable quantities, and also quadratic irrationalities. In modern times the formula was extended to $c^2 = a_1^2 + \cdots + a_n^2$ for the diagonal of a rectangular parallelepiped in n-dimensional Euclidean space, and gave the expression

$$s^2 = (x^1 - y^1)^2 + \cdots + (x^n - y^n)^2 \qquad (1)$$

for the distance of two points $x = (x^1, \cdots, x^n)$, $y = (y^1, \cdots, y^n)$ in Cartesian coordinates there. For infinitesimally near points this becomes the line element $ds^2 =$

$(dx^1)^2 + \cdots + (dx^n)^2$ which in the more general affine version

$$ds^2 = \sum_{i,j=1}^{n} g_{ij} \, dx^i \, dx^j \tag{2}$$

is the cornerstone of *Riemannian geometry*. On the other hand, analytic developments demanded the extension of (1) to a space of infinitely many dimensions and then even to the case in which coordinates are complex numbers, thus

$$s^2 = |z^1 - w^1|^2 + |z^2 - w^2|^2 + \cdots$$

With this definition of distance the space becomes a Hilbert space, basis of operator theory and quantum mechanics. Finally, again for finite dimension but still complex coordinates, it became significant, in analogy to (2), to introduce the line element

$$ds^2 = \sum_{\alpha,\beta=1}^{n} g_{\alpha\bar{\beta}} \, dz^\alpha \, dz^{\bar{\beta}}$$

It is the basis for the study of Riemannian geometry on *complex manifolds,* and the study of such manifolds is penetrating even into Relativity and Field theory.

The harmless relation $(1 + x)^2 = 1 + 2x + x^2$ was generalized by Newton to

$$(1 + x)^n = 1 + \binom{n}{1} x + \binom{n}{2} x^2 + \cdots \tag{3}$$

The coefficients are the *binomial coefficients*

$$\binom{n}{k} = \frac{n(n-1)\ldots(n-k+1)}{1 \cdot 2 \cdot \ldots \cdot k} \tag{4}$$

and the formula is a *generating relation* for them. Expression (4) can also be set up if n is any noninteger number, real or complex, and it was indeed established that (3) holds for any n for $|x| < 1$. The expansion is usually an infinite series, and it is in fact the Taylor series of $(1 + x)^n$. By studying, for general n, the behavior of the series as $|x| \to 1$, H. Abel laid the foundation for the theories of summability of non-convergent series.

Euclid's theorem that any integer is a unique product of primes (stated in Euclid only implicitly) expresses itself as a formula

$$a = p_1^{n_1} \cdot p_2^{n_2} \cdot \ldots \cdot p_k^{n_k}$$

The concept of a prime factor and the quest for a theorem on unique decomposition into prime factors has been a leading principle of arithmetic and algebra on all levels ever since.

Thus, although our algebra had begun during the Renaissance, long before the Calculus, it was fully inaugurated only during the second half of the 19th century, when E. Kummer and R. Dedekind constructed so-called "ideals," with which to maintain the theorem of unique decomposition into prime elements, even for elements of algebraic number fields. After this construction had proved successful, algebra introduced other basic concepts, such as field, ring, extension, matrix algebra, etc., with which to reassess its storehouse of formulas and formulaic verities, and to carve out a major novel division of contemporary mathematics entirely for itself.

The constant π, widely known from antiquity through the formulas $l = 2\pi r$, $A = \pi r^2$ for a circle and $A = 4\pi r^2$, $V = \frac{4}{3}\pi r^3$ for a sphere, also enters the sophisticated formula

$$\frac{\pi}{4} = 1 - \frac{1}{3} + \frac{1}{5} - \frac{1}{7} + \cdots$$

due to Leibniz. Both kinds of formulas were much used for the approximate computation of this *transcendental num-*

ber, and a good deal of computational techniques (including computing machines) was initiated in the course.

The transcendental number *e,* basis of natural logarithms, appears in the 18th century in the formulas

$$e^{2\pi\sqrt{-1}} = 1, \qquad \cos \phi + \sqrt{-1} \sin \phi = e^{\phi\sqrt{-1}}$$

which have been of immeasurable consequence to all of mathematics. Even the age-old plane trigonometry was greatly streamlined by them.

Euler (end of 18th century) gave for convex polyhedra the formula $e_0 - e_1 + e_2 = 2$ in which e_0 is the number of vertices, e_1 of edges and e_2 of surfaces. This was a first step in function theory of closed Riemann surfaces and in algebraic geometry, and it was a starting point for H. Poincaré's combinatorial topology. Poincaré's own formula $B_p = B_{n-p}$ for Betti numbers of compact manifolds is even today one of the most arresting statements of the theory.

Cauchy's formula

$$f(z) = \frac{1}{2\pi i} \int_C \frac{f(\zeta)d\zeta}{\zeta - z}$$

is a centerpoint from which the theory of complex functions radiates.

Galileo's main achievement is embodied in the formula $s = \frac{1}{2}gt^2$ for a falling body and Newton's gravitational law in the formula $F = m_1 m_2 / r^2$, which also represents Coulomb's law of electricity. A. Einstein's daring formula $E = mc^2$ from special relativity created the atomic bomb, and Heisenberg's interpretation (1926) of the formula $pq - qp = h/2\pi i$ from quantum theory as an *indeterminacy relation* for physical observables has been an irritant to philosophers ever since.

3. FOUNDATIONS: MATHEMATICAL LOGIC

A prime demand on mathematics is that it should be deductively rigorous, and a traditional model for intended rigor is Euclid's presentation of mathematical assertions in Theorems. For the great philosopher G. Spinoza in the 17th century, "more geometrico" is a synonym for "deductively rigorous."

A theorem is a proposition which has been proved, excepting certain first theorems, called axioms, which are admitted without proof; and to prove a theorem means to obtain it from other theorems by certain procedures of "deduction" or "inference." It had been a commonplace for long that each branch of mathematics is based on its own axioms, but the 19th century arrived at the insight that even the same branch might have alternate axioms. Specifically, there were envisaged alternate versions of two- and three-dimensional geometry, the axiom varied being the axiom on parallels (non-Euclidean geometries). It was also recognized that a set of axioms becomes mathematically "possible" if it is logically "consistent," that is, if one cannot deduce from the axioms two theorems one of which, as a proposition, is the negation of the other. At the same time, certain developments led to the realization that, not only the axioms, but the rules of inference themselves might be, and even ought to be, subject to variations. Now if axioms and rules of inferences are both viewed as subject to change it is customary to speak of a *mathematical system* or also a *formal system,* and, of course, an irreducible first requirement is this that the system shall be consistent after the manner just stated. Consistency alone is a rather negative property. There is a further property, called *completeness,* which is rather more positive, and which, if present, is very welcome. A system is complete if for any proposition which can be formulated it either can be proved that it holds or that its negation holds. If a theorem holds in a system and if the system is altered then the same proposition, or what corresponds to it in the new system, may become undecidable if not outright false; and if it remains valid it may require a new proof because certain axioms or rules of inference are no longer available.

Some of the developments which led to doubt whether the traditional rules of inference are inviolate were the following:

(1) G. Boole had found (1854) that the classical Aristotelian connectives "and," "or," "negation of" for propositions follow rules similar to those which the operations "addition," "multiplication," "the negative of" obey in ordinary algebra (Boole's algebra of propositions), and this gradually took from rules of inference the priviledged status of untouchability.

(2) George Cantor, the founder of the theory of sets and operations between them, defined a set ("intuitively" or "naïvely") as the collection of "all" objects having a certain property which is verbally expressible. Especially "the set of all sets" is again a set and it has the pecularity of being a set which contains itself as one of its elements. But this leads to the following contradictory situation (Russel paradox). Divide the totality of all possible sets into two categories. A set shall belong to category I if it does not contain itself as an element, and to category II if it does contain itself as an element. Now form the set M whose elements are the sets of category I. It can now be reasoned by deductive steps admissible in Cantor's own theory that the set M cannot belong to either of the two categories, although the original division into categories did assign each set to one of them.

(3) In 1904, E. Zermelo formulated the following "axiom of choice." Given any family of non-vacuous sets (S), no matter how (infinitely) large the family may be, it is possible to choose "simultaneously" an element $x = x_S$ from each given set S and thus to consider the set M consisting of precisely these elements. Now, by the use of this axiom some striking theorems in classical mathematics could be proved which, without the use of the axiom,

seemed to be logically out of reach entirely. Mathematicians began to wonder whether a theorem based on the axiom of choice is indeed valid or, at any rate, whether it has the same level of validity as one without it, and as a consequence of this, theorems employing the axiom of choice were being frequently labeled as such.

Some of the doubts were resolved eventually, and the most striking results to date are the following ones of Kurt Gödel:

(1) Any consistent mathematical system which is sufficient for classical arithmetic must be incomplete. In other words given any formal system expressible in the arithmetical language taught to a student of a secondary school in any country it is possible to formulate assertions which can be neither proved nor disproved.

(2) Any such system remains consistent if one adds to it the axiom of choice, so that working mathematics cannot disprove the axiom of choice. (In 1963, P. Cohen showed that working mathematics remains also consistent if one adds the negation of the axiom of choice, so that the latter axiom cannot be proved either.)

(3) The so-called general continuum hypothesis (which we will describe immediately) is also consistent with ordinary mathematics, and, in fact, ordinary mathematics remains consistent if the axiom of choice and the general Continuum hypothesis are added simultaneously.

Cantor says that two sets A_1, A_2 have the same *cardinal number* if there exists a one-to-one correspondence between the elements of A_1 and A_2. Any infinite set has the same cardinal number as some proper subset of itself. For instance the set of all integers $A_1 = (1, 2, 3, \ldots)$ has the same cardinal number as the subset of all even integers $A_2 = (2, 4, 6, 8, \ldots)$ because the association of the number n in A_1 to the number $2n$ in A_2 is one-to-one. However for a finite set this cannot happen, and the cardinal number of a finite set is the "ordinary" number of its elements. Next, Cantor says that a set A_2 has a larger cardinal number than A_1 if there is a one-to-one correspondence from A_1 to a subset of A_2 but not one from A_2 to A_1 or a subset of A_1. He then shows that if A is any set, and S is the set of its subsets including the empty set, then the cardinal number of S is greater than that of A. If A is finite and has n elements, then S has 2^n elements. Now 2^n is indeed greater than n, and for $n \geqq 2$ there are even other integer numbers between n and 2^n. This is so for finite sets. But for infinite sets the situation seems to change radically, and the "generalized continuum hypothesis" asserts that there are no other cardinal numbers between those of A and S.

4. COUNTEREXAMPLES

It follows easily from the definition of a derivative that a differentiable function $y = f(x)$ in $a < x < b$ is also continuous. Now, K. Weierstrass deepened considerably this statement by demonstrating that the converse is not true. He constructed a function $f(x)$ which is continuous in $a < x < b$ but does not have a derivative at any point (but his function does have right and left derivatives at all points). A construction of this kind is called a *counterexample*. Systematic preoccupation with counterexamples, which sometimes require considerable ingenuity, did not arise before the middle of the 19th century. Another early counterexample of considerable consequence referred to Fourier series. If $f(x)$ is periodic with period 2π then its Fourier series is

$$\frac{1}{2} a_0 + \sum_{n=1}^{\infty} (a_n \cos nx + b_n \sin nx) \tag{5}$$

where, by definition,

$$a_n = \frac{1}{\pi} \int_0^{2\pi} f(t) \cos nt \, dt, \quad b_n = \frac{1}{\pi} \int_0^{2\pi} f(t) \sin nt \, dt$$

Now, Fourier himself stated somewhat vaguely that for "any" function $f(x)$ the partial sums of the Fourier series are covergent to $f(x)$. Dirichlet was the first to give a specific criterion of convergence. In his criterion the function $f(x)$ may have certain simple discontinuities. But in the present discussion we exclude such and we demand that $f(x)$ shall be continuous, first of all.

Now, P. G. Lejeune Dirichlet proved that the Fourier series converges everywhere to $f(x)$, if $f(x)$ is piecewise monotone, that is, if the interval of periodicity $0 \leqq x < 2\pi$ may be divided into a finite number of smaller intervals such that in each of these the function is either monotonely increasing or monotonely decreasing. P. du Bois Reymond demonstrated that this requirement of monotoneity cannot be dispensed with entirely. He constructed a continuous periodic function $f(x)$ for which the Fourier series fails to converge in at least one point. By a general principle due to A. Harnack ("condensation of singularities") it is then possible to construct a continuous function whose Fourier series fails to converge at countably many points even if they are everywhere dense, for instance at all rational points $x = p/q$. But to this day it is not known whether there exists a continuous function whose Fourier series fails to converge at all points, or at least at all points with the only exception of a set of Lebesgue measure 0.

Counterexamples frequently fall into patterns, meaning that the same underlying construction is used over again for different but related purposes. A construction used in many counterexamples is Cantor's *ternary set*. Divide the closed interval $0 \leqq x \leqq 1$ into three equal parts $(0, \frac{1}{3})$, $(\frac{1}{3}, \frac{2}{3})$, $(\frac{2}{3}, 1)$ and remove the middle one, but only the *open* inner part of it $\frac{1}{3} < x < \frac{2}{3}$. This leaves two closed intervals $0 \leqq x \leqq \dfrac{1}{3}$, $\dfrac{2}{3} \leqq x \leqq 1$. In each of these remove the *open* inner third, that is $\dfrac{1}{9} < x < \dfrac{2}{9}$ and $\dfrac{2}{3} + \dfrac{1}{9} < x < \dfrac{2}{3} + \dfrac{2}{9}$. This leaves four closed intervals altogether. From each remove the open inner third, etc. After n steps the number of intervals removed is $1 + 2 + 2^2 + \cdots + 2^{n-1}$ and the sum of the removed intervals is

$$\frac{1}{3} + \frac{2}{3^2} + \frac{2^2}{3^3} + \cdots + \frac{2^{n-1}}{3^n} = 1 - \left(\frac{2}{3}\right)^n$$

This value is less than 1, which must so be, since the total interval has length 1 and the removed intervals do not overlap. However if we let n go to ∞ then $\left(\dfrac{2}{3}\right)^n$ tends to 0, and the aggregate length of all infinitely many intervals which have been removed has value 1. Thus the pointset which is left over ought somehow to have a one-dimensional measure which has value 0. This is so indeed in the theory of Lebesgue. Also the leftover set is "nowhere dense." However it is a very big set still, by a crude "counting" of its points at any rate. More precisely, it turns out that its cardinal number, as defined in section 3, is the same as for the entire interval $0 \leqq x \leqq 1$. This can be deduced from the following fact, which is interesting for many purposes. Any number $0 \leqq x \leqq 1$ can be represented instead of by a decimal expansion, by a ternary expansion, that is, by an expansion

$$0 \cdot a_1 a_2 a_3 \ldots$$

in which each symbol a_1, a_2, a_3, . . . has one of values 0, 1, 2. Now, our leftover set consists of precisely those points in whose expansion the symbols a_1, a_2, a_3, . . . assume only the values 0, 2 and not 1.

5. CONSTRUCTIVENESS: APPROXIMATIONS

Some mathematicians object to "mere existence proofs," and they demand that any proof shall also be "constructive." The interpretations of this demand differ widely. Some of them come rather near to what a "practical" mathematician welcomes; if for instance a theorem asserts the existence of a number or a function, then the proof must also embody a procedure for actual computation of the "solution," approximately at least. Other versions are little more than the negative requirement that certain combinations of inference shall be avoided. And there are also views which combine both, and the best known among the last is the *intuitionist* view. It firmly demands a certain kind of constructiveness, which however does not necessarily guarantee the calculation by present-day computing machines, say. But the actual stricture by which Intuitionism became widely known is this, that proof-by-contradiction is not admissible. It is also called proof by double negation, and it is equivalent to the Aristotelian "tertium non datur." A proof-by-contradiction assumes tentatively that the proposition to be proved is false and from this assumption deduces a contradiction to a theorem previously established.

There is one difference between practical and theoretical demands on constructiveness. Sometimes it is possible to prove the existence of a function non-constructively and after that, on the basis of the existence statement, to devise a procedure for a manageable construction. To practical mathematics this is satisfactory, to theoretical mathematics it is not.

A demand superficially related to constructiveness, and without any overtones of "logic," is a desire to estimate the degree of approximation, whenever a task of approximation is being performed. For instance, a classical result of Lagrange for the "Taylor series with remainder term" states that if $f(x)$ has n derivatives in the interval $0 \leqq x \leqq 1$, then $f(x) = P_n(x) + R_n(x)$ where $P_n(x)$ is the polynomial

$$P_n(x) = \sum_{m=0}^{n} \frac{f^{(m)}(0)}{m!} x^m \qquad (7)$$

and the "error term" $R_n(x)$ is in numerical value bounded by $M_n/n!$, where M_n is the maximum of $|f^{(n)}(x)|$ in $0 \leqq x \leqq 1$.

This result leads us to the following topic. The polynomial (7) can only be formed if $f(x)$ has n derivatives. However, there is a theorem of Weierstrass that any function $f(x)$ which is continuous in an interval $a \leqq x \leqq b$ (and nothing more) can be approximated uniformly by polynomials

$$P_n(x) = a_0 x^n + a_i x^{n-1} + \cdots + a_n \qquad (8)$$

Each proof for this theorem produces its polynomials and there is an extensive theory for securing those for which the approximation is a best possible one.

The parallel question of approximating not by ordinary polynomials (8) but by trigonometric polynomials

$$s_n(x) = \frac{1}{2} a_0 + \sum_{m=1}^{n} (a_m \cos mx + b_m \sin mx) \qquad (9)$$

leads to mathematical procedures which pervade all of science nowadays. First, if one wishes to make the difference

$$\max_{0 \leq x \leq 2\pi} |f(x) - s_n(x)| \qquad (10)$$

rather small, then the partial sums of the Fourier series (5) are not very good, meaning that it is not very advantageous to use the polynomials (9) in which the coefficients a_m, b_m

are the expressions (6). However it improves the degree of approximation if one repalces the partial sums by their arithmetic means (Fejer sums)

$$\sigma_n = \frac{s_0 + s_1 + \cdots + s_{n-1}}{n}$$

which are again exponential polynomials, namely

$$\sigma_n(x) = \frac{1}{2} a_0 + \sum_{m=1}^{n} \left(1 - \frac{m}{n}\right)(a_m \cos mx + b_m \sin mx)$$

There are other "averages" which approximate even better, and one is led to studying procedures of "averaging and smoothing" which are of importance to Analysis, Probability, and Statistics.

But a major development begins if, following Bessel and Parseval (beginning 19th century), we replace the expression (10) by the expression

$$\left(\frac{1}{2\pi} \int_0^{2\pi} |f(x) - s_n(x)|^2 \, dx\right)^{1/2} \tag{11}$$

as a measure for the degree of approximation. In this case the partial sums of the Fourier series are indeed the best approximating sums for any order n, not only for ordinary Fourier series but for so-called orthogonal series in general. Pertinent application to the theory of accoustical and then optical waves led to an interpretation of the Fourier series (5) as a "spectral resolution" of its function $f(x)$, and this interpretation has been gradually extended to a host of related and analogous expansions in Analysis, Algebra, Theory of Probability, Quantum Theory, and other parts of Physics. The measuring of "approximation," "deviation," "dispersion," etc. by a square of integral mean—which is (11)—has become a strong and pervasive influence on scientific thinking throughout.

6. SPACE IN MATHEMATICS

If geometry is the mathematics of space then, in a superficial sense, all mathematics began with geometry, since, apparently, it began with measurements attaching to figures: length, area, volume, size of angles. However it did not concern itself with questions of shape but with clarifying and deciding when figures are equal or substantially equal with regard to form. The first true theory of Geometry was the great theory of the Greeks. It saw its main task in studying the basic concept of equality of figures in the two versions of *congruence* and *similarity*, and it was so determined to dissociate itself from the preceding phase of making nothing but measurements that Euclid's extensive book, for instance, avoids any kind of actual measurements to a fault. In Euclid's proof of the Pythagorean theorem $c^2 = a^2 + b^2$, there is not a breath of a hint that it might be an equality between numbers. It is only an equality between areas based on congruences, two of the squares being cut up into pieces and then put together into the third square after the manner of a jigsaw puzzle. But, for all its lofty purposes, Greek geometry was too rigid and circumscribed to be able really to cope with the mathematical problem of space. It ran itself out dead, and geometry did not move until, with the advent of Coordinate Systems in Geometry, a true mathematics of Space could initiate.

If we etch a Cartesian coordinate system into two- or three-dimensional Euclidean space, this makes the space into a pointset, each point being a pair (x^1, x^2) or a triple (x^1, x^2, x^3) of real numbers, and any figure a suitable subset of it. This is a deliberate process of *Arithmetization of Space* which unifies space and number at the base. This does not hamper geometry in its task of pursuing problems of shape but, on the contrary, aids it. In the Cartesian plane, two figures are similar if the points of

[270]

one can be obtained from the points of the other by means of a transformation

$$y^1 = a^1 + \alpha_1{}^1 x^1 + \alpha_2{}^1 x^2, \quad y^2 = a^2 + \alpha_1{}^2 x^1 + \alpha_2{}^2 x^2 \quad (12)$$

where

$$\alpha_1{}^1 \cdot \alpha_1{}^2 + \alpha_2{}^1 \cdot \alpha_2{}^2 = 0 \quad (13)$$

and for some $\rho > 0$

$$(\alpha_1{}^1)^2 + (\alpha_2{}^1)^2 = (\alpha_1{}^2)^2 + (\alpha_2{}^2)^2 = \rho^2 \quad (14)$$

The similarity is a congruence if and only if $\rho = 1$ (orthogonal transformation). Now this analytic representation of congruence and similarity suggests a geometric examination of the most general linear transformations (12) which are non-singular, that is, for which the determinant $|a_q{}^p|$ is $\neq 0$. They were virtually unknown to the Greeks, although they highlight the axiom of parallels of Euclid's geometry. A one-to-one transformation of the Cartesian plane is such a linear transformation if and only if it carries a straight line into a straight line and parallel straight lines into parallel straight lines. Thus a parallelogram goes into a parallelogram; and, in fact, given any two parallelograms there is a linear transformation which carries the one into the other no matter what the angles and ratios of sides in the two figures are. There is a geometry, the so-called affine geometry, in which any two parallelograms are considered "equal." It cannot measure angles, and non-parallel segments cannot be compared as to length. However this geometry does have conics, and it can separate them into ellipses, parabolas, hyperbolas.

The family of all linear transformations constitute a so-called *transitive group,* and the subfamily of orthogonal transformations is already a transitive group. Now, Felix Klein has made the pronouncement, which is generally accepted, that there arises a geometry on a space if on the space there is given a transitive group of transformations, and if two figures are considered equal whenever one figure can be carried into the other figure by one of the transformations. For the non-Euclidean geometries of Bolyai–Lobatschewsky various "models" have been exhibited which conform to this view, and perhaps the most interesting one is the following. In the plane (x^1, x^2) introduce the complex variable $z = x^1 + ix^2$ and consider the family of transformations

$$w = e^{i\theta} \frac{z - a}{1 - az}$$

for all constant complex numbers a for which $|a| < 1$ and all real numbers θ. They give rise to the following "non-Euclidean" geometry. The "space" is not the entire z-plane but only the unit disk: $|z| < 1$. A "point" is an ordinary point in it, and a "straight line" of the new geometry is in the Euclidean geometry either a diameter of the disk or any circular arc inside the disk whose endpoints are on the boundary line of the disk and which meets the boundary line of the disk at right angles. Through a point outside a straight line of the geometry there go infinitely many straight lines which are "parallel" to it, that is, do not meet it inside the disk. This makes the geometry a "hyperbolic" one. If conversely in a geometry any two straight lines always intersect, the geometry is called "elliptic." The prototype of an elliptic geometry is a geometry on a surface of a sphere (say, the surface of the earth) in which the straight lines are the great circles. As a matter of fact, the hyperbolic geometry just described is in a very precise sense a counterpart to the elliptic geometry of great circles on a sphere.

The arithmetization of space led to a purely mathematical creation of n-dimensional space, Euclidean and other, for any integer dimension n, by defining its points generally as n-tuples of real numbers $(x^1, \ldots, x^n)$ with suitable definitions for various "geometrical" relations between such points. The most eye-catching consequence in science was the four-dimensional space of the theory of relativity, but, as we have already stated in Chapter 5, multi-dimensional geometry had, in a sense, been playing a part in physics before that. If a mechanical system involves M mass points it was customary in effect to consider the space of dimension $n = 3M$ whose points are the states of the system, that is the n-tuples of coordinates (x_m^1, x_m^2, x_m^3), $m = 1, \ldots, M$, at any one time point. Also if there are restraints operative in the system, then the Lagrange–Hamilton theory suitably reduces the dimension of the space by the use of the "free parameters" of the system instead of the original n coordinates themselves. The use of free parameters spread from mechanical systems to other systems in physics, chemistry, etc., and all so-called equations of state are geared to such. Finally, in quantum theory a "state" of a system has infinitely many coordinates, and the infinitely dimensional space representing it is a Hilbert space. Also, partly under the influence of Hilbert space, mathematicians have become fascinated with infinitely dimensional spaces in general. They are being studied intensively, and large parts of mathematics are being pressed, not always gently, into these new frames of reference.

Also, the arithmetization of space reflects itself in the ever widening use of graphs and charts to many purposes. Any tabulated dependence of a number y on a number x is a function $y = f(x)$ and hence representable as a curve on graph paper, and a great amount of information is thus illustrated and stored. We even "chart" a course of action nowadays. Furthermore the glove compartment of any automobile is filled with road maps, that is, local charts, and any larger area is covered by the use of several such local charts which suitably overlap. This is nothing else but a practical, day-in-day-out application of the concept of a "manifold" in modern topology and differential geometry, and, as frequently, the "abstract" formulation of a mathematical concept is but a circumlocution of what "common sense" dictates.

[273]

CHAPTER 8

THE ESSENCE OF ANALYSIS*

ANALYSIS is one of three main divisions of mathematics, the other two being (i) Geometry and Topology and (ii) Algebra and Arithmetic. In extent, Analysis is the largest; it comprises subdivisions which are nearly autonomous and which are easier to describe than the whole division.

I. MEANING OF ANALYSIS

1. RELATION TO SCIENCE

Greek mathematics had a geometrical tenor, as the works of Euclid, Archimedes, Apollonius of Perga, and Pappus testify. It did initiate some durable topics of analysis, but the organized creation of analysis began only in "modern" times around A.D. 1600. Analysis then came into being in stages, growing up in intimacy with Mechanics and Theoretical Physics. This is not to say that other mathematics does not enter into science too, in fact, all mathematics does. The two great innovations in physics in the 20th century, Relativity and Quantum Theory, had to rely heavily on existing mathematical tools of geometric and algebraic provenance. But whenever basic physics instigated a topic in mathematics spontaneously it was largely in analysis. Thus, differential and integral calculus, ordinary differential equations, and calculus of variations have arisen from mechanics; Fourier series from acoustics and thermodynamics; complex analysis from acoustics, hydrodynamics, and electricity; partial differential equations from elasticity, hydrodynamics, and electrodynamics; and even mathematical probability, which falls under analysis,

* Reprinted from *Encyclopaedia Britannica*, 1961–. Slightly revised; originally entitled "Analysis."

although born from problems of gambling and human chance, drew much of its syllogistic strength in the 19th century from statistical theories of mechanics and thermodynamics.

2. DERIVATIVES

Geometry deals with spaces and configurations, topology with spatial deformations, algebra with the general nature of the basic operations of addition, subtraction, multiplication, and division, and arithmetic with additive and multiplicative properties of general "integers." Now, analysis also deals with specific operations, namely with differentiation and integration. But it is more appropriate to say that it deals with the mathematical "infinite" in many of its aspects, as: infinite multitude, infinitely large, infinitely small, infinitely near, infinitely subdivisible, etc. Its first objects and concepts—introductory and yet actively basic —are: infinite sequence, infinite series, a function $y = f(x)$, continuity of a function, derivative of a function $y' = df/dx$, and integral of a function. (The words "derivative," "function," "integral" in the English language all appeared in the 16th century; as names in Analysis, "derivative" goes back to G. W. Leibniz in 1676 and "function" in 1692, and "integral" to J. Bernoulli in 1690.) The mathematical concept of derivative is a master concept, one of the most creative concepts in analysis and also in human cognition altogether. Without it there would be no velocity or acceleration or momentum, no density of mass or electric charge or any other density, no gradient of a potential and hence no concept of potential in any part of physics, no wave equation; no mechanics, no physics, no technology, nothing.

The formal textbook definition of the concept took over 150 years to evolve, but even to the untutored it will be rewarding to savor Isaac Newton's own description of it under the name of "ultimate ratio" (we quote from page

39 of Florian Cajori, Sir Isaac Newton's Mathematical Principles, University of California Press, 1934).

> For those ultimate ratios with which quantities vanish are not truly the ratios of ultimate quantities, but limits towards which the ratios of quantities, decreasing without a limit, do always converge: and to which they approach nearer than by any given difference, but never go beyond, nor in effect attain to, until the quantities have diminished *in infinitum.*

Newton's generation was literate enough to listen to a recondite mathematical definition, all in words, without symbols; and the Marquise du Châtelet, woman of the world, could undertake to translate Newton's formidable *Principia* from Latin into French, ultimate ratios and all.

The semantic structure of the quoted sentence was well within reach of Greek natural philosophy. Similar sentences can be found in Euclid and Archimedes, and, outside of mathematics, as early as Thucydides. Now, if the inspiration of Greek thinking had been such as to be able to bring about the cognitive content of this sentence as well, then undoubtedly Aristotle would have devoted much of his "metaphysics" to it or even written a special treatise "On the Art of Derivation," and Archimedes might have discovered the 17th century mathematics and physics which he had been dimly perceiving all his life.

3. REAL NUMBERS

Mathematical formulation of scientific statements introduces a peculiar kind of lucidity and precision into them and it suggests and establishes logical and cognitive relations between them. It also introduces challenging analogies and unifications. For instance, most phenomena of propagations of waves, whether in acoustics, or in hydrodynamics, electricity, or optics are assumed to be governed at the outset by virtually the same set of differential equations. Thus for a while the theories run parallel, until the actual differences of aim must be begun to be accounted for.

The most basic trait of unification is the fact that in the last "analysis" all specific and tangible information in science is quantitative, that is, expressed in ordinary real numbers. Every physical magnitude, whatever its own unit of "dimension" (simple or composite) may be, is measured by one or perhaps several real numbers, and all these numbers come from the same joint supply. Without this fact, there would be no Constancy Laws in physics, such as the law of the constancy of energy is; it could not be enunciated or even conceived. Space extensions, time intervals, noises, colors, tactile impressions, and emotions, all these things are in the end "valued" quantitatively. The meaning in depth of formulas, equalities, and inequalities in science comes to life by accompanying interpretations; but the transmittable specific information is all in numbers, whether on graphs or on punch cards or on tape.

This all-pervading arithmetization in physics and also mathematics was systematically begun in the 17th century, and in mathematics its manifest outcome was first of all the "analytic" or "coordinate" geometry. Mathematics is less insistent than physics on its exclusiveness. On the contrary, 20th-century mathematics abounds and delights in ever introducing and rarely discarding concepts and objects which need not be numerical. But, one somehow frequently comes back to numbers, and the effect of the 17th-century arithmetization of space on the course of mathematics cannot be over-stated.

4. MAGNITUDES

We already said in Chapter 6 that the Greeks did not have real numbers but, in its place a notion of "magnitude" = $\mu\acute{\epsilon}\gamma\epsilon\theta\sigma\varsigma$. In Homer this noun still means: personal greatness or stature (of a hero, say); and it is remarkable that for instance in the French noun *grandeur* and the German noun *Grösse* the two meanings of personal greatness and of mathematical magnitude likewise reside simultaneously. Twentieth-century mathematics has "magnitudes" on various levels of generality, but it has them in addition to and not instead of real numbers. The Greek attempt of by-passing real numbers altogether was amazingly mature, but in retrospection it appears to have been *too* mature. The Greeks took a giant stride, stopped, and advanced no further.

II. STAGES OF DEVELOPMENT

5. IRRATIONAL NUMBERS

In the 5th and 4th centuries B.C. the Greeks gave the first known satisfactory solution to a problem in analysis. In a square whose side has length 1, the diagonal has length $\sqrt{2}$. The decimal expansion of $\sqrt{2}$ is the non-terminating symbol 1.4142 . . . , so that $\sqrt{2}$ is the sum of the infinite series

$$1 + \frac{4}{10} + \frac{1}{100} + \frac{4}{1000} + \frac{2}{10,000} + \cdots$$

Now, the Greeks proved that $\sqrt{2}$ is not a rational number, that is, not a quotient p/q for integers, p, q. But they apparently found for $\sqrt{2}$ an approximate representation by certain rational numbers, or rather they found for the diagonal of the square approximations by segments which are commensurate with the side of the square. The approximate numbers were what were later called the partial sums of the continued fraction for $\sqrt{2}$, and apparently it was the study of this problem from which evolved the so-called Euclidean Algorithm. The deepest achievement in this entire context was the development of a general theory of incommensurate ratios of certain types of "magnitudes" and of proportions between ratios. The fifth Book of Euclid which contains this theory was much admired by the scholar and statesman Sir Isaac Barrow, predecessor to Newton in the Lucasian Professorship at Cambridge University, from which post Barrow resigned in favor of Newton in order to devote himself to state craft entirely.

6. AREAS AND VOLUMES

The Greeks also investigated the number

$$\pi = 3.14159 \ldots$$

which is the area of a circle of radius 1, and the problem of "squaring the circle" was so widely known to the Athenian public that Aristophanes could allude to it in his comedy *Birds* (414 B.C.). Computations of other areas and volumes and of centers of masses are known from the works of Archimedes, and these results taken cumulatively can be viewed as a precursor to the 17th-century theory of Integration. But these "direct" Greek calculations of many particular areas and volumes by circumscribed and inscribed approximating figures, when compared with the procedures of the Integral Calculus proper, are heavy, tiring, and tiresome.

7. SET THEORY

The Greeks were also puzzled by the fact that the ordinary number 1 is the sum of the geometric series

$$\frac{1}{2} + \frac{1}{4} + \frac{1}{8} + \frac{1}{16} + \cdots + \frac{1}{2^n} + \cdots \tag{1}$$

and this was for them part of a range of problems relating to "infinitely small" and "infinitely large." The preoccupation with such problems is reflected in the paradox of "Achilles and the Turtle" put forward by Zeno of Elea and in the distinction between "potential" and "actual" infinite as proposed by Aristotle. Partly in order to refute this distinction, the 19th-century mathematician Georg Cantor developed his *Set Theory* or *Theory of Aggregates* the first beginnings of which go back even to Galileo. This theory not only strengthened the logical framework of analysis, but soon became a basic innovation in all of mathematics, and in the speculations about its essence and its general background.

8. ANALYTIC GEOMETRY

The Greek achievements in analysis were but a prelude. A notable step was taken in the 14th century by Bishop Nicolas Oresme. He discovered the logical equivalence between tabulation and graphing, more or less, and in a way he proposed the use of a graph for plotting a variable magnitude whose value depends on that of another. However the systematic theoretical basis for this possibility evolved only later from the work of (Pierre Fermat and) René Descartes, who in 1637 laid the foundation for Analytic Geometry, that is, a geometry in which everything is reduced to numbers. In this geometry a point is a set of numbers which are called its coordinates. A figure is viewed as an aggregate of points, but it is usually described by formulas, equations, and inequalities. This approach to geometry is the very basis for maps, graphs, and charts which have penetrated into all walks of life, and for the mathematical concept of a function $y = f(x)$ which, first intended only for analysis, has eventually penetrated into all corners of mathematics equally, and into all areas of rational thinking as well.

Later on in the 17th century the culminating event took place when Newton and Leibniz introduced derivatives and laid the foundation for calculus and mechanics. Archimedes had been uneasily groping for the concept of a derivative, especially in his book *On Spirals,* but he could never hit upon it, although otherwise he would have been as well equipped for dealing with it as anybody in the 16th and 17th century was. The 17th and 18th centuries produced much mathematics in all directions but analysis was the underlying theme. The titles of the two great works of J. L. Lagrange carried the words, Mécanique "analytique" (1788) and Théorie des fonctions "analytiques" (1797).

9. 19TH CENTURY

In the 19th century, topology, algebra, and arithmetic were emerging as independent divisions of mathematics, rivaling analysis in attention, but analysis continued to proliferate and, by sheer bulk, to predominate. Many large French treatises on general mathematics simply bore the title "Cours d'analyse." But, by a peculiarity of usage, in memory of the fact that Euclid's *Elements* was always viewed as a book in geometry, in these same treatises a mathematician in general was frequently referred to as "un geomètre," especially when he was "illustre."

10. 20TH CENTURY

Only in the 20th century were topology and algebra beginning to assert a status approximating that of analysis itself. At the same time the lines of demarcations between the main divisions of mathematics were beginning to be flexible and uncertain. Since the 1930's, analysis has been penetrating into geometry and arithmetic farther than ever, on the one hand, and, on the other hand, topology and algebra have been beginning to impose tighter and more streamlined structural schemes on the large body of analysis, much to her benefit. Duplications and redundancies could be reduced by syllogistic identifications and assimilations, and in some areas an increase of the substance offered itself naturally.

Furthermore, digital computing machines are injecting

into analysis a certain algebraization, if only marginally. The structure of the machines and the problem of coding both involve algebra of logic, and in computational assignments from analysis, limiting processes must be replaced by approximations within certain maximal numbers of steps.

III. FUNCTIONS

11. HIGHER DERIVATIVES

A function $y = f(x)$ for real numbers x, y is equivalent with a graph in an (x, y)-coordinate system. A graph looks like a curve, but a curve is a figure on ordinary paper whereas a graph is a figure on graph paper; and graph paper represents a Euclidean plane into which some fixed coordinate system has been introduced beforehand.

The derivative df/dx, at a value x, is the limit of the difference quotient $\Delta f/\Delta x$ as Δx tends to 0. It is defined for the numerical function but has an interpretation for the graph as a figure, namely as the slope of the straight line which is tangent to the figure at the given point with abscissa x. However, the numerical value of the slope does depend on the coordinate system given. This dependence becomes more emphatic when one introduces the second derivative of $f(x)$, that is, the derivative of the derivative

$$\frac{d^2f}{dx^2} = \frac{d}{dx}\left(\frac{df}{dx}\right)$$

In order to form it one must view the first derivative itself as a new function—no matter how inarticulately or deficiently at first—and then form the derivative of this new function in turn. Second derivatives were at the center of Newton's mechanics. They defined the kinematic concept of acceleration as the instantaneous rate of change of the velocity function, and then, by an interpretative correlation with it, the dynamic concept of force. The Greeks had no functions or accelerations, and only varying notions of force.

12. GENERAL FUNCTIONS

A comprehensive concept of function was beginning to emerge only in the first half of the 19th century, at first only for numbers x, y and gradually for more general objects x, y. In the case of general objects x, y one also uses instead of the word "function" the words "functional" or "operator," especially when functions for various types of objects are used in the same context simultaneously. For *working* mathematics, that is, when problems relating to *foundations* of mathematics are not at issue, the following unsophisticated description of a function $y = f(x)$ in general seems sufficient.

There are given two aggregates of mathematical objects, an aggregate X and an aggregate Y. Let x denote an element of X and y an element of Y. With each element x of X there is associated a certain element y of Y, and a single one only. But the same element y may be associated to several different elements x, although it is not necessary that each element y shall appear as an associate to some element x. For instance, it is admissible that *all* elements x have as associates the same element y, in which case the function is called a constant. In general the element y associated to x is denoted by a symbol like $y = f(x)$. It is also called the image of x, or also the value of f at the "point" x. The entire object $y = f(x)$ is called a function from X into Y or, more explicitly, from the aggregate X into the aggregate Y.

Syllogistically a mathematical statement is a certain conclusion drawn from particular assumptions and previous statements. A function in general, as just described, is not sufficiently tangible to start making statements about, even if X and Y are real numbers (except if one of them is a finite set). Rather, in order to start, it is necessary to impose on the functions some restrictions or qualifications, that is, to single out classes of functions with some particular features to analyze. Together with the general notion of function there also began emerging certain over-all "descriptive" properties by which to single out classes of functions, as for instance: continuity, differentiability, integrability, bounded variation, etc. But before that, especially in the 17th century it was taken for granted, expressly or not, that a function is a "formula" or "expression" that is a tangible prescription for finding y if x is given. Certain functions, then familiar, were taken as known, being basic, as it were, as for instance $x, x^2, \ldots,$ sin x, tan x, arctan x, e^x, log x; and a formula arose by taking one or several such functions and performing a finite number of "natural" operations, such being primarily the four arithmetical basic operations, but also extraction of roots and, what is important, the substitution of one function into another.

However soon there appeared in the manipulations certain expressions in which an infinite, that is unending, succession of operations was involved, and which were very suggestive too. Such were certain infinite series, infinite products, and infinite Continued Fractions. Of these the infinite series proved the most important ones by far, and two types of such series have advanced into positions of leadership and command which no phase of development has seriously impaired since; although there have been cases of evasion from authority, half in mathematical sport and a little in earnest. These two are: power series and Fourier series.

13. POWER SERIES

The simplest power series is the geometric series

$$\frac{1}{1-x} = 1 + x^1 + x^2 + x^3 + \cdots$$

and in a broad sense it is very "ancient," since for instance Zeno's series (1) is a case of it for $x = \frac{1}{2}$, if one subtracts the number 1 from both sides. The general power series has the form

$$a_0 + a_1 x^1 + a_2 x^2 + \cdots + a_n x^n + \cdots \tag{2}$$

in which a_0, a_1, a_2, . . . are fixed numbers and x is a variable which for the present we assume to be real-valued. Now, if such a series is convergent for all x in an interval $-R < x < R$, where R is some positive number, and if one denotes the sum by $f(x)$, then this function has "very good" properties. It is continuous, even differentiable, and in fact it has a second derivative, third derivative, etc. Furthermore if one knows the values of all those derivatives at one point, namely at the origin $x = 0$, then the coefficients a_n of the series (2) can be computed. They are

$$a_0 = f(0),\ a_1 = \frac{1}{1} f'(0),\ a_2 = \frac{1}{1.2} f''(0), \ldots,$$

$$a_n = \frac{1}{1 \cdot 2 \ldots n} f^{(n)}(0)$$

[Brook Taylor, 1715]. Also, it is possible to shift the origin $x = 0$ into any other fixed point $x = h$ by starting out with a series

$$a_0 + a_1(x - h) + a_2(x - h)^2 + \cdots + a_n(x - h)^n + \cdots$$

If this series converges in some interval $x_0 < x < x_1$, which contains the "special" point h, say in an interval $-R + h < x < R + h$, and if one denotes the sum function again by $f(x)$, then it has derivatives of all orders in the interval, and moreover

$$a_0 = f(h),\ a_1 = \frac{1}{1} f'(h), \ldots,\ a_n = \frac{1}{1 \cdot 2 \ldots n} f^{(n)}(h), \ldots$$

$$\tag{3}$$

Now, in order to compute all these derivatives only for the point h, it suffices to know the function $f(x)$ in some part interval containing the point h, no matter how small this part interval may be, and the coefficients a_0, a_1, a_2, . . . , are already determined thereby. If therefore we consider the class of functions $f(x)$ which can be represented as sums of power series, then we obtain the following *uniqueness property*. If two such functions are equal in some small interval, then they are equal in any larger interval containing it.

The 18th century was very impressed with this property because it agreed with a scientific presumption that if a closed mechanical system has been known for a certain time interval then this knowledge ought to determine it for later time as well. Lagrange especially claimed, with more insistence perhaps than self-conviction, that mechanics, physics, and even "significant" analysis in general ought to be able to do with such functions exclusively. He firmly attached to these functions the name "analytic" (probably already current before) which they have borne since, and in his works cited in section 8 he tried to use them "exclusively."

14. PIECEWISE ANALYTIC FUNCTIONS

However this "exclusively" still allowed for a certain generalization which Lagrange (like others) was taking for granted due to the following situation. If one throws a ball against a wall and it hits it at the time t_0, then in the time interval $0 \leq t \leq t_0$ the components of position and velocity and other quantities for the ball are analytic functions of the time variable and of each other. After striking the wall at the time point t_0, the ball flies on its rebound during a certain time interval $t_0 \leq t \leq t_1$. During the rebound the functions are again analytic, as they were before, but at the instant of impact itself this is not so. For instance at $t = t_0$ the velocity components are not even continuous, let alone analytic. Such "instantaneous" interruptions of analyticity, when the "instants" were only finitely many, were acceptable to Lagrange and others. Their occurrence was taken for granted, and viewed as a necessary generalization of the demand on analyticity if not of its basic definition.

In the 20th century such functions were called "piecewise" analytic, and this manner of generalization became important in many contexts. There are functions which are piecewise continuous, piecewise differentiable, etc. An indispensable method in algebraic topology, which was initiated in 1911 by L. E. J. Brouwer, is built on approximating to continuous functions in several variables by piecewise linear ones.

15. FOURIER SERIES

Physics and technology views many of its phenomena as being "waves" (under which "vibrations" and "oscillations" also fall), and there are waves in acoustics, hydrodynamics, electricity, optics, etc. In each context there is a certain first type of them, not yet too complicated and yet already important, and the associated interpretations and names are closely analogous from context to context. There is always a certain decomposition of the general wave within that type into a series of component waves, which themselves are no longer decomposable. The first in the series is the "basic," "lowest," "primary," etc.; the later ones are "refinements," "higher," "secondary," etc., and they are supposed to decrease in effect and influence the later they come in the series. Also each of these components has certain fixed attributes, such as (wave) length, frequency, phase, amplitude, etc.

Now, this analogy in the "phenomena" is simply due to the identity of the mathematical instrument used for their "decomposition." Invariably, a certain function of one variable x in $-\infty < x < \infty$ is being represented as a so-called Fourier Series, and the attributes of the phenomena are simply attributes of its series.

This function is then a periodic function, that is, there is

[288]

a fixed number p such that $f(x + p) = f(x)$. We are normalizing it by putting $p = 2\pi$ so that $f(x + 2\pi) = f(x)$, and its Fourier Series is then an expansion

$$f(x) = \tfrac{1}{2}a_0 + (a_1 \cos x + b_1 \sin x)$$
$$+ (a_2 \cos 2x + b_2 \sin 2x) + \cdots$$
$$+ (a_n \cos nx + b_n \sin nx) + \cdots \quad (4)$$

in which a_0, a_1, a_2, . . . ; b_1, b_2, . . . are certain constants. The first coefficient is written with the coefficients $\tfrac{1}{2}$ in order to make the following formulas uniform

$$a_n = \frac{1}{\pi} \int_{-\pi}^{\pi} f(x) \cos nx \, dx, \quad b_n = \frac{1}{\pi} \int_{-\pi}^{\pi} f(x) \sin nx \, dx \quad (5)$$

All terms in the series (4) have period 2π, however for the computation of the coefficients a_n, b_n by the formulas (5) it suffices to know the function $f(x)$ only in the interval $-\pi < x \leqq \pi$. This agrees with the fact that if a function $f(x)$ is given in the latter interval to start with, then it can be extended to a periodic function by repeating it successively in the intervals $(\pi, 3\pi)$, $(3\pi, 5\pi)$, . . . to the right and similarly in the intervals $(-3\pi, -\pi)$, $(-5\pi, -3\pi)$, . . . to the left.

For purposes of theory and of physics it is appropriate to rewrite the series (4) into two more forms. The first is the "complex" version

$$f(x) = c_0 + (c_1 e^{ix} + c_{-1} e^{-ix}) + \cdots$$
$$+ (c_n e^{nix} + c_{-n} e^{-nix}) + \cdots$$

which arises by the substitution

$$\cos nx = \frac{e^{inx} + e^{-inx}}{2}, \quad \sin nx = \frac{e^{inx} - e^{-inx}}{2i}$$

and in this version

$$c_m = \frac{1}{2\pi} \int_{-\pi}^{\pi} f(x) e^{-imx} \, dx$$

for all integers m, positive and negative. The second form arises if, for $n \geqq 1$, we introduce the number

$$\rho_n = (a_n{}^2 + b_n{}^2)^{1/2}$$

and an angle θ_n for which

$$a_n = \rho_n \cos \theta_n, \qquad b_n = \rho_n \sin \theta_n$$

The series then becomes

$$f(x) = \tfrac{1}{2} a_0 + \rho_1 \cos (x - \theta_1) + \rho_2 \cos (2x - \theta_2) + \cdots$$
$$+ \rho_n \cos (nx - \theta_n) + \cdots$$

In physical applications it is frequently pertinent to replace x by $x - at$ where t is time and a is the velocity of the wave, and the result is

$$f(x - at) = \tfrac{1}{2} a_0 + \rho_1 \cos (x - at - \theta_1) + \cdots$$
$$+ \rho_n \cos (nx - nat - \theta_n) + \cdots$$

The constant term $\tfrac{1}{2}a_0$ does not represent anything "wavy" and it is frequently 0. However the next term

$$f_1(x - at) = \rho_1 \cos (x - at - \theta_1)$$

represents the lowest indecomposable component of the wave, and the successive terms

$$f_n(x - at) = \rho_n \cos (nx - nat - \theta_n) \tag{6}$$

the higher ones. Also $2\pi/n$ is the length of (6), $na/2\pi$ its frequency, θ_n its phase and ρ_n its amplitude. The interpretation of the amplitude attaches itself to an important relation, called the Parseval equality. It states that for any Fourier series (4) one has

$$\frac{1}{\pi} \int_{-\pi}^{\pi} (f(x))^2 \, dx$$
$$= a_0{}^2 + (a_1{}^2 + b_1{}^2) + \cdots + (a_n{}^2 + b_n{}^2) + \cdots$$
$$= a_0{}^2 + \rho_1{}^2 + \rho_2{}^2 + \cdots + \rho_n{}^2 + \cdots$$

For some wave phenomena the quantity

$$\frac{1}{\pi} \int_{-\pi}^{\pi} (f(x))^2 \, dx$$

represents a certain volume of energy associated with the propagation of the wave. Now, ρ_n^2 is the volume of energy for the component wave $f_n(x)$. Therefore, if one "ignores" the neutral term a_0 then the Parseval equality states that this energy for the whole wave is the sum of the energies for all component waves.

16. ANALYTIC THEORY OF NUMBERS

The trigonometric expansion (4) is named after J. Fourier largely because he found a counterpart to it in which the sum is replaced by an integral. It is the pair of formulas

$$g(y) = \int_{-\infty}^{\infty} f(x)e^{2\pi iyx} \, dx$$

$$f(x) = \int_{-\infty}^{\infty} g(y)e^{-2\pi ixy} \, dy \tag{7}$$

and they belong together in the following sense. For certain classes of functions $f(x)$ in $-\infty < x < \infty$ the first integral exists for all y in $-\infty < y < \infty$, so that a function $g(y)$ arises. With this function one can form the second integral (7), and the resulting function $f(x)$ is the one from which the process started. This pair of formulas has given rise to a theory of "duality" in Abstract Analysis.

S. D. Poisson, and others before him, gave a formula which combines the Fourier series with the Fourier integrals, and the simplest case of it is

$$\sum_{m=-\infty}^{\infty} \int_{-\infty}^{\infty} f(x)e^{2\pi imx} \, dx = \sum_{n=-\infty}^{\infty} f(n)$$

It became one of the most important tools in the Analytic Theory of Numbers when P. G. Lejeune Dirichlet used it in 1837 for giving a new proof for the so-called Gauss Reciprocity Law. In number theory, the greatest master of this so-called Poisson Summation Formula was Erich Hecke (1887–1947).

17. LEAST SQUARE APPROXIMATION

There is an amplification to the Parseval theorem which has shaped many a conception in physics.

Let $f(x)$ be a function in $-\pi < x \leqq \pi$. For a fixed integer $n \geqq 1$ take any real numbers $A_0, A_1, \ldots, A_n,$ $B_1, \ldots, B_n$ and set up the trigonometric sum

$$(8) \quad \sigma_n(x) = \tfrac{1}{2} A_0 + (A_1 \cos x + B_1 \sin x) + \cdots + (A_n \cos nx + B_n \sin nx)$$

A particular case of this sum is the n-th partial sum

$$s_n(x) = \tfrac{1}{2} a_0 + (a_1 \cos x + b_1 \sin x) + \cdots + (a_n \cos nx + b_n \sin nx)$$

of the Fourier series (4), that is, it is among the sums $\sigma_n(x)$ the one which arises if one chooses for the coefficients A_m, B_m the numbers a_m, b_m which are given by the formulas (5). But now form the difference $f(x) - \sigma_n(x)$ and ask for which choice of the coefficients A_m, B_m this difference will be a "least possible one." A good answer to this question arises if one measures the size of the difference by a certain "average" of it, namely by the value of the integral

$$\frac{1}{\pi} \int_{-\pi}^{\pi} |f(x) - \sigma_n(x)|^2 \, dx \qquad (9)$$

This quantity is always greater or at best equal to

$$\frac{1}{\pi} \int_{-\pi}^{\pi} |f(x) - s_n(x)|^2 \, dx \qquad (10)$$

so that it is smallest if the trigonometric sum (8) is the corresponding partial sum of the Fourier series for $f(x)$. Furthermore, the function $f(x) - s_n(x)$ has itself a Fourier series which arises by cutting off from the entire series (4) the partial sum $s_n(x)$, thus being

$$a_{n+1} \cos (n + 1)x + b_{n+1} \sin (n + 1)x + \cdots$$

Therefore, by Parseval equality, the integral (10) has the value

$$a_{n+1}^2 + b_{n+1}^2 + a_{n+2}^2 + b_{n+2}^2 + \cdots \tag{11}$$

and altogether the following result arises. If the degree of approximation is measured by the integral (9), and if $\sigma_n(x)$ is a trigonometric polynomial of prescribed degree n, then the best approximation is brought about by the partial sum of the Fourier series and its size is (11).

This result is not restricted to expansions in which the "simple" functions are $(\cos nx, \sin nx)$ but it holds if they are a so-called complete ortho-normal system, and in this way it broadens out into a comprehensive proposition in many settings.

18. FUNCTIONS OF REAL VARIABLES

The uniqueness property for analytic functions (see section 13 above) is indicative of great regularity of structure. Given a small piece of the function, that is, given the function in a small interval, it thence acquires a natural growth into a larger interval and uniquely so. Also if one places the x-line on which it is defined as an x-axis into the plane of the complex variable $z = x + iy$, then it even spreads in a unique manner from the one-dimensional interval on the x-axis into a two-dimensional domain of the plane. Syllogistically the decisive phenomenon of this spread is the fact that if a power series (2) is convergent on a real interval $-R < x < R$, and if one replaces the real variable x by the complex variable $z = x + iy$ then the power series with the same coefficients

$$a_0 + a_1 z^2 + a_2 z^2 + \cdots$$

will also converge in the entire disk $|x + iy| < R$.

The complex analytic functions thus arising give rise to a theory of great structural variety and beauty. The mathematicians of the 19th century loved and admired it all. They gave it their best, and molded into it all the glamour of the "Victorian era" then regnant everywhere and in everything. Around 1900 and later a number of books on such functions bore the generic title "Theory of functions" in the expectation that the "other" functions would identify themselves expressly; and so they did by calling themselves functions of a "real" variable or of real variables.

The theory of the latter functions somehow was also growing into a compound of many parts, and in the 20th century it was even beginning to gain an ascendancy of interest over the complex analytic area. But then the complex analytic functions started to revitalize themselves by branching out from one into several variables, and a balance of evenly divided interest was establishing itself.

Fourier series are intimately related to the Theory of Functions of real variables but not to all of it, and yet they were a great ferment for the development of the theory in all parts. Functions of bounded variation arose from a paper by Dirichlet giving systematically the first nonobvious converge criterion for Fourier series. B. Riemann developed the theory of his Integral in a paper on trigonometric series, and G. Cantor, in pursuing a problem raised in that paper, was led to constructing the first part of his *theory of sets,* namely the theory of pointsets in Euclidean Space. The theory of summability of divergent series, originally designed as an aid for analytic continuation of power series, became lastingly indispensable when L. Fejèr applied it to Fourier series. The Lebesgue integral was recognized in its importance when it led to the Riesz–Fischer theorem for Fourier series and to the Plancherel theorem for Fourier integrals; and the Stieltjes integral, originally designed for the so-called moment problem, acquired its stature when it entered Fourier Analysis in the theory of Probability.

IV. ANALYSIS AND SPACE

19. CONFORMAL MAPPING

In the plane of Cartesian coordinates (x, y) one can introduce the *polar coordinates*

$$\rho = (x^2 + y^2)^{1/2}, \qquad \theta = \arctan \frac{y}{x} \qquad (12')$$

in the following way. On the half-line which issues at the point of origin $(0, 0)$ and passes through the point (x, y), the quantity ρ is the distance between the two points and θ is the angle between the positive half of the x-axis and this half-line. From ρ, θ the numbers x, y can be re-obtained by

$$x = \rho \cos \theta, \qquad y = \rho \sin \theta \qquad (12'')$$

There are some complications in the use of (ρ, θ). First the origin itself, called the *pole,* is a singularity in the sense that no specific θ can be assigned to it, and secondly, even omitting the pole, the angle θ cannot be assigned continuously to the entire plane, but only for any sector of the plane, say, the positive quadrant $x > 0$, $y > 0$. These difficulties are familiar from the use of latitude and longitude on the earth. The earth is spherical, not planar, and due to that feature there are two poles, at which no longitude is definable; and if one starts measuring longitude continuously eastwards, say from Greenwich, then on returning to it the accumulated longitude is 360° and not 0° as at the start.

Coming back to the relations $(12')$, $(12'')$, we replace

the letters ρ, θ by "neutral" letters u, v, and rewrite these relations thus

$$u = \phi(x, y), \qquad v = \psi(x, y)$$
$$x = A(x, y), \qquad y = B(x, y) \qquad (13)$$

There are many other possibilities for choosing functions ϕ, ψ together with A, B; and each choice creates a so-called curvilinear coordinate system of which polar coordinates are a particular instance. In three-dimensional space many kinds of curvilinear coordinate systems are useful to mechanics and physics.

The pair of functions (13) can also be interpreted entirely differently. In addition to the given (x, y)-plane we introduce another Euclidean plane with coordinates u, v, this one also Cartesian and we view (13) as a mapping which associates with each point $P : (x, y)$ of the first plane a point $Q : (u, v)$ of the second plane. Such a mapping falls under the general concept of function as formulated in section 12, and the word "mapping" is meant to recall a map in geography which arises when the points of a region of the earth are "mapped" into a certain coordinate system on a sheet of paper or on an artificial globe.

Two particular types of mapping are dominant in the geometry of Euclid, congruences and similarities. In traditional geometry two triangles are congruent if, in modern parlance, there is a certain one-to-one mapping of the entire Cartesian plane into itself which carries one triangle into the other. Euclid's geometry characterizes these mappings by certain axiomatic requirements, but in terms of coordinate transformations they are

$$u = a + \alpha x + \beta y, \qquad u = b + \gamma x + \delta y$$

with certain properties, which we have described in Chapter 7.

Such analytic formulations of congruence and similarity suggest the introduction of more general transformations, and prominent among them is *conformal mapping,* which is defined as follows. If one takes any point (x_0, y_0) and all "small" triangles having this point as one of the vertices then, in the limit, their images are similar to them, but the size of the local dilation may vary with the point (x_0, y_0); that is, conformal mapping preserves angles but not distances. Two statements are indicative of the importance of such mappings. First, if one also demands that the sense of the angles be preserved then conformal transformations are the same as complex analytic functions in the following manner. If a pair of functions (13) constitutes a conformal mapping, and if one introduces the complex variables $z = x + iy$, $w = u + iv$, then there is a complex analytic function $w = f(z)$ such that $f(z) = \phi(x, y) + i\psi(x, y)$. Conversely for any complex analytic function, its real and imaginary parts constitute a sense-preserving conformal transformation. Secondly, if one takes in the plane a region bounded by one curve only, then it can be mapped conformally into the interior of a circle, and this also holds if the region is on a sphere, or on any two-dimensional surface in space. Thus, for any plot of land on the earth which can be enclosed by one continuous fence one can make a conformal (that is angle-preserving) mapping into the interior of a circle on a flat sheet of paper. The oldest conformal mapping is the Stereographic Projection of Hipparchus, and of Ptolemy (ca. A.D. 150), and its name is due to F. Aguilonius (1613).

20. DEGREES OF FREEDOM

A point P in space has three coordinates (x, y, z). If there are given N points P_n, $n = 1, \ldots, N$, if the coordinates of the n-th point are denoted by (x_n, y_n, z_n), and if the points constitute a mechanical system in motion, then at any given time the system is described by the $3N$ numbers (x_n, y_n, z_n), $n = 1, 2, \ldots, N$. Now, there may be "constraints" operative in the system. For instance, two of the points, say $P_1 : (x_1, y_1, z_1)$ and $P_2 : (x_2, y_2, z_2)$ may be linked by a rigid rod of length l so that always

$$(x_1 - x_2)^2 + (y_1 - y_2)^2 + (z_1 - z_2)^2 = l^2$$

In this case it is "economical" to eliminate one of the coordinates, say by putting

$$x_1 = x_2 + \left(l^2 - (y_1 - y_2)^2 - (z_1 - z_2)^2\right)^{1/2}$$

so that only $3N - 1$ "free" parameters are left. The physicists then say that the "true" number of degrees of freedom for the system is (at most) $3N - 1$, and it is sometimes preferable to interpret the motion of the system in a space of (at most) $3N - 1$ dimensions. If there are two "restraints" present, and if they are "independent" then there are (at most) $3N - 2$ degrees of freedom, etc. In Mechanics, the theory of degrees of freedom was initiated by Lagrange, and the mathematical aspects of the phenomenon of "dependence" between functions, and the subsequent reduction of the total dimension, was put on a sure footing by C. G. J. Jacobi (1804–1851) and A. Cayley (1821–1895). These studies gradually led to the emergence of a precise concept of dimension and of (topological) manifold. The general public became aware of such possibilities at a rather advanced stage of the development, namely in the 1920's by the Theory of Relativity of Albert Einstein, which featured two interlocking hypotheses. It merged the three-dimensional "ordinary" space with the time variable into a physically inseparable four-dimensional space, and it allowed this total space to be a topological manifold which was no longer a Euclidean space in its entirety.

Algebraic topology itself was especially stimulated by a formula due to L. Kronecker (1823–1893), the so-called Kronecker integral, which represents the number of joint solutions of three independent functions

$$u = f(x, y, z), \qquad u = g(x, y, z), \qquad w = h(x, y, z)$$

[298]

in a region of three-dimensional Euclidean space, and similarly for higher dimensional space. The formula subsumes the so-called Cauchy formula in Complex Analysis and related formulas which are more familiar than Kronecker's formula itself.

21. TENSORIAL FUNCTIONS

Let a differentiable function $g(u, v)$ denote the height of a building erected over a domain D in the (u, v)-plane as base. If one introduces a transformation of coordinates (13) then the height becomes

$$f(x, y) = g(\phi(x, y), \psi(x, y))$$

Such a function is called an *absolute scalar*. Now, the "skyline" of the building, that is, the shape of its roof is appropriately described by the two functions

$$\frac{\partial g}{\partial u}, \quad \frac{\partial g}{\partial v} \tag{14}$$

In the coordinates x, y the corresponding functions

$$\frac{\partial f}{\partial x}, \quad \frac{\partial f}{\partial y} \tag{15}$$

are linked to the functions (14) by the relations

$$\frac{\partial f}{\partial x} = \frac{\partial g}{\partial u} \frac{\partial \phi}{\partial x} + \frac{\partial g}{\partial v} \frac{\partial \psi}{\partial x}$$

$$\frac{\partial f}{\partial y} = \frac{\partial g}{\partial u} \frac{\partial \phi}{\partial y} + \frac{\partial g}{\partial v} \frac{\partial \psi}{\partial y}$$

so that in order to obtain either one of the two functions (15) both functions (14) must be known. In this sense the pairs (14), (15) cannot be broken up into singles. In general, over manifolds with local coordinate systems, there are many types of groups of functions which must be kept together organically for purposes of finding their values under changes of coordinates. Such groupings of functions are called *tensors,* and the name first arose in Elasticity to describe "tensors" whose components are strains and stresses in different directions. In the topology of manifolds, tensorial functions are vector bundles, more or less.

A very important tensor is the *curvature tensor* introduced by Riemann and anticipated by Gauss. It has inaugurated a new era in geometry, and although its adequate description is rather technical, it is a very unifying concept nonetheless. It brings together geometry proper, analysis, topology, and even a branch of Analytic Theory of Numbers (automorphic functions in several variables). Also, there would hardly be a General Theory of Relativity without it.

PART II. BIOGRAPHICAL SKETCHES

BIOGRAPHICAL SKETCHES

AESCHYLUS (525–456 B.C.), Attic Tragedian, the first, and still one of the greatest, tragedians in the "West" of whom entire plays survive. Aeschylus derived from landed aristocracy, and, although a lifelong poet by avocation, he led, as customary with Athenians of his generation, a crowded citizen's life in the service of the polis. For instance, tradition has it, and there is no reason to doubt it, that he was present at the battles of Marathon, Artemisium, Salamis, and Plataea in the years 490–479.

Of his few surviving works, his last was his trilogy *Oresteia* (*Agamemnon, Choeophoroi, Eumenides*) which was presented in 454, only two years before his death. There are Greek scholars who view the *Oresteia* as the greatest poetical work of Greece, ahead of the *Iliad*.

The leading achievement of Aeschylus was the introduction of the dramatic dialogue, and his dialogue is immortally exemplified in the *Agamemnon*, especially in the encounter between Klytaemnestra and Cassandra. Before Aeschylus, in Homer say, even when persons spoke passionately in rapid succession to one another, it still sounded as if they were observing some rules of parliamentary conduct in town meetings and war councils; only Aeschylus created the poignancy which goes with a stage dialogue as we have it today.

The language in Aeschylus, although contrived, is superb, and it does not lose its force in translation. But Aeschylus cannot be performed on our stage today because his "Greek" chorus still has an inordinately large part in the proceedings and in the unfolding of the plot, and this is fatiguing both to stage directors and to audiences.

Aeschylus is religiously motivated, and his Godhead is

Zeus; also, even an evil woman like Klytaemnestra is somehow still a somewhat-more-than-human "heroine." It has been noted—and there are even a few special books about it—that the plaints of Prometheus against Zeus in the *Prometheus Chained*—and some analogous features in other plays of Aeschylus—are reminiscent of the plaints of Job against Jahwe in the Old Testament, and that the author of *Job* may have been synchronous with Aeschylus. But the similarity must not be pressed too far. Nor, on the opposite side, must one attempt to find that the texture of the religious spirit in Aeschylus is such, as to pre-announce the forthcoming prevalence of philosophical and scientific rationality, a century hence. No amount of literary criticism or analysis of Shakespeare would elicit from his work a pre-announcement of the composition of Newton's *Principia* less than a century after him.

AGUILONIUS, F. or FRANÇOIS D'AGUILLON (1567–1617), Jesuit from Brussels, taught philosophy at Douai, and theology at Antwerp at a college of which he was rector. The expression *stereographic projection* appears in his treatise on optics, 1613.

ALHAZEN (965–1039), born at Bassora; died in Cairo, after feigning insanity for many years in order to escape death for not being able to regulate the inundations of the Nile as he had offered to do. He composed many works in mathematics, astronomy, medicine, philosophy, and physics, most of them lost. His treatise on optics, which became known to Kepler through Witelo, was the greatest single work on this topic between Ptolemy and Kepler.

AMONTONS, GUILLAUME (Paris 1663–id. 1705), deaf from childhood; physicist, worked successfully in friction and thermometry. In a sense, he anticipated the Gay-Lussac law, in memoirs beginning with one in 1699.

AMPÈRE, ANDRÉ MARIE (Lyons 1775–Marseille 1836). His father died on the guillotine in 1793. He was a front-rank French physicist, father of electrodynamics, creator of the conception of a "current" of electricity instead of Oersted's "conflict" of electricity. Before turning to physics he taught, and earned reputations in, mathematics, philosophy, and chemistry; and he left behind an unfinished *Essai sur la philosophie des sciences,* which was intended to be a grandiose synthesis of all scientific knowledge. He occupied at various times many high academic posts; but at his death, which took place while he was on an inspection tour of universities, he was half-forgotten.

ANATOLIUS (SAINT) (Alexandria 230–Laodicea A.D. 282), bishop of Laodicea since 276, Aristotelian, one of the most educated persons of his times; presumed author of certain fragments dealing with the nature of mathematics, author of a work on the date of Easter.

ANAXAGORAS OF CLAZOMENAE (500–428 B.C.), died at Lampsacus; spent most of his life in Athens where he was the first resident philosopher, and where his book was on sale, for a drachma, at the entrance to the theatre.

To the man in the street he was probably known for the assertion that the celestial bodies are but red hot stones; apparently because of that, although reputed to be a gentleman of pure character, he had to stand trial for blasphemy in 434. To the thoughtful he offered the philosopheme that there are occurrences in nature, on a cosmic scale, which are out of the ordinary, and which can only be explained by the intervention of Intelligence Personified (= Nous). He had a theory of the constitution of matter with which Aristotle had his difficulties. Anaxagoras maintained that there are infinitely many substances which are qualitatively different, but that "there is a portion of everything in anything," meaning that corpuscles can interpenetrate, after the manner of de Broglie waves of matter in our contemporary physics. Aristotle had an inkling of this meaning when he changed the terms *spermata* (seeds) and *chremata* (things = elements = particles) of Anaxagoras into *homoiomeré* (having [spatially] similar parts).

Aristotle, for purposes of his own, emphasized the views of Anaxagoras on the microscopic composition of matter. Some commentators after Aristotle followed suit, and this created the background against which commentators of today are anxious to overcome apparent inconsistencies in extant fragments of Anaxagoras relating to this topic. However, Anaxagoras himself was much more a many-sided natural philosopher than a specialist in physics proper. Like "everybody else" he did have to have views on the "particle structure" of matter to begin with. But he was broadly interested in cosmology and astronomy; in "Life Sciences"; in the working of the senses; and, perhaps more than in anything else, in the problem of the relation between Mind and Matter, to which problem, after all, there has hardly been any kind of answer since. It is not justified to assume, as sometimes done, that, for Anaxagoras, these various topics coalesced into one, with the topic of "particle physics" being in the center of it all; especially, since Plato and Aristotle, with the book of Anaxagoras before them, could not make out what his Nous (-Mind) was meant to be or to achieve.

ANAXIMANDER OF MILETUS (610–545 B.C.), younger contemporary of Thales, was a fascinating person; someone with whom to talk for hours about anything and everything. Yet Plato never mentions him, just as he never mentions Democritus, who was equally fascinating, and who was even nearer to his times and concerns.

There are reports that Anaximander discovered the

obliquity of the zodiac; that he installed devices for the determination of hours, or of seasons; that he constructed a "sphere," apparently a celestial one; and that he sketched a world map, which, in a more detailed version, was afterwards used by the world traveler Hecataeus, perhaps as an appendix to a book on his travels. Furthermore, Anaximander taught that the sun and the moon are each a circular band of fire which is encased in a diurnally rotating envelope of cosmic mist. The envelope is invisible, except for an aperture through which the sun or the moon shines forth; and a darkening or an eclipse arises if the aperture narrows or closes. There are similar statements, somewhat vague ones, about fixed stars or planets. Anaximander also had some views, not very clear ones, on biology: man evolved from an aquatic creature or, at least, was reared inside a piscine animal until he was ready to fend for himself.

However, in a sense all this is only peripheral. Anaximander's main achievement, and the *only* one of which Aristotle takes notice, is Anaximander's attempt to formulate a physico-metaphysical conception of something which he called *apeiron* (= infinite, illimited, unbounded, undetermined, undefinable, etc.) and which apparently was meant to be some kind of "intangible" originative cosmic substance, self-identical even if subject to modifications. It is not known to what extent Anaximander was able to articulate a description of his *apeiron,* and there is much to suggest that his notion of it was pre-articulate and indistinct. Apparently with reference to modifications of his *apeiron,* Anaximander is supposed to have made the pronouncement, which was handed down by Theophrastus, that "growth and decay happen from necessity, in payment and retribution for unusual injustices, by the assessment of Time." Textbooks notwithstanding, this pronouncement is unintelligible, and Aristotle ignores it. Instead, Aristotle undertook to explicate the *apeiron* as if it were an ancestor to the physico-ontological notion of *Infinitude* which was crowding in on his thinking from all parts; and his essays on "infinite," "place," "void," and "time" in Books III and IV of the *Physica* led eventually to the construction of transfiniteness by Georg Cantor. Also, Aristotle knew that his infinite was not really the original *apeiron,* and he quoted Anaximander rarely and cautiously.

ANAXIMENES OF MILETUS (c. 588–524 B.C.), came after Thales and Anaximander. According to Aristotle and Doxographers he taught that all substances arise from air [$=\dot{\alpha}\dot{\eta}\zeta$] through rarification (fire) or condensation (water, earth, stones); that the Earth is a flat disk floating stably on air; and that, in earthquakes, mountains crumble either when excessively dry or when excessively moist.

The physics of Anaximenes is unimaginative. Also, there are no enlivening anecdotes about him. Diogenes Laertius dismisses him in a few lines, and says, in effect, that his style of writing was uninteresting.

ANNA COMNENA (1083–1118), famed Byzantine lady historian; she was versed in Greek writings beginning with Homer, and herself wrote in a stiltedly archaizing but competent Greek.

She was the daughter of emperor Alexius I. Comnenus, and after his death tried to have her brother John supplanted on the throne by her husband Nicephorus Bryennius, himself a historian. The plot failed, and, retiring from politics, Anna turned to literary activity. Her work, the *Alexiad,* is a panegyrical family history, about events from the period 1069–1118, which was part of her father's reign. The period embraces the First Crusade, which, for Anna, was a threat to the Byzantine empire. Of particular interest in her work are her reactions to the Crusade and to individual crusaders.

APOLLONIUS OF PERGA, studied in Alexandria, and there are indications that his treatise in *Conics* was composed around 200 B.C.

The *Conics* created the genre of an exhaustive monograph on a topic in mathematics; it pre-required a certain "standard" level of mathematics in general, except that in the topic to which it is devoted it begins entirely from the beginning. The work is in eight "books," that is "rolls," and there are most enlightening epistolary introductions to most of them. The introduction to Book I contains, among others, two points which could occur in a book of today. First, the author warns the reader that there are some "pre-publication" versions of the work in circulation, for which he must not be held academically responsible. Secondly he announces that "Volume 1" consisting of Books I–IV, deals with "more elementary" properties of conics, whereas "Volume 2" consisting of Books V–VIII is more for the specialist in quest of research topics.

Regrettably, in consequence of this statement by the author himself, many libraries apparently did not acquire Volume 2; at any rate, Book VIII, that is, the last "chapter" in Volume 2, is altogether lost, and of Books V–VII only Arabic translations have come down.

There is yet a historical problem to be solved. From the manner in which Kepler quotes Apollonius, it is all but certain that he was well familiar with the content of Book V. But if he was, the question arises how this familiarity came about; because the first known translation of the Arabic text into Latin was printed in 1661, long after Kepler.

ARCHIMEDES OF SYRACUSE (c. 287–212 B.C.), son of an astronomer, was Greece's star mathematician. By avocation he desired the pursuit of mathematics proper, and he was wholly and passionately committed to mathematics at its "purest." But by world reputation he was an engineer, especially in the field of military engines, even if he protested that he derived no satisfaction from this kind of work. And when confronted with the problem of determining whether a golden crown was made of pure gold or was alloyed with silver, he initiated the method of Hydrostatics for the purpose.

Scientist-professors will be always the same. Archimedes, when heading the Weapons Research Group for the Syracuse Department of Defense, would write letters to friends that he was yearning to return to the Campus and do nothing but pure research for its own sake. But he was apparently doing classified work to his last breath, literally so. And when he found his theorem in Hydrostatics he was so excited that he insisted on talking about it to the man in the street.

From the manner in which he mentions them, Archimedes must have thought highly of the astronomer Aristarchus of Samos, and of the mathematician and astronomer Eudoxus of Cnidos, and also of the philosopher Democritus of Abdera, whom however he mentions, once, qua mathematician only. And it is moving to see with what skill Archimedes manages, in the *Sand-Reckoner,* to mention the name of his own father, Phidias, obviously an astronomer, in one breath with that of the renowned Eudoxus.

Archimedes is on friendly terms with his near-contemporary Eratosthenes of Cyrene, of Sieve-of-Eratosthenes repute, whose main claim to fame is that of having been the first really scientific geographer of Greece; and he treats as colleagues mathematicians Dositheus, Heraclides, and Zeuxippus, about whom, apparently, nothing is ascertainable.

Also, Archimedes was, for years, demonstratively disconsolate over the early death of his friend Conon of Samos, of whose mathematical potentialities Archimedes speaks in highest terms. Conon must have been indeed promising, because Apollonius in his preface to Book IV of the *Conics* mentions him favorably, although agreeing with one Nicoteles of Cyrene that some of the proofs of Conon were not what one would wish for. Also, Pappus states vaguely, in Book IV, Chapter 21, of his *Collection,* that Conon did some initial work on the spirals of Archimedes; other reports have it that Conon even made a creditable finding in astronomy. For a summary about Conon, see Paul Tannery, *Mémoires scientifiques,* III, 353–354.

ARISTOTLE (384–322 B.C.) was a polymath, the greatest ever, with equal emphasis on "poly" and "math"; and his polymathy was the greatest single organizing force in the establishment of Greek, and hence Western, rationality. Plato, with Socrates before him, may have raised most of the searching philosophical problems that are. However, it is not enough only to raise problems; and it is a fact that frequently Aristotle dared to articulate answers, however questionable ones, where Plato had only dared to ask, or to equivocate, or to let the problem lapse into abeyance.

Aristotle was born in the little town of Stagira, and he is *the* Stagirite; at the age of 17 he "enrolled" in Plato's Academy, when Plato was already 61, and remained in it for 20 years, till the age of 37, when Plato died. The 12 years between 347 and 335 he spent away from Athens, in part as a tutor to Alexander the Great. In 334, when Alexander acceded to the throne, Aristotle returned to Athens, where he stayed, as master of his own school, till 323, when Alexander died. Feeling insecure in Athens he then withdrew to Chalcis, but died soon afterwards, in 322, at the age of only 62.—Many efforts notwithstanding, it is difficult to establish some kind of chronology of Aristotle's extant works. For instance, it is impossible to say at which stage or stages of his life his *Physica* was composed.

In his personal life, Aristotle was orderly, conformist, and "normal." He was not very impressive of physique, so he took care to be always well groomed. He was kind to his slaves, faithful to his friends, and he took good care of his family and of his near relatives. His testament was well thought out, and was apparently kept up to date.

Aristotle was born into a tradition of medicine, and thought patterns of biology are dominant in his work. About a third of his extant work is in "life-sciences," including psychology; and Charles Darwin stated, more or less, that Aristotle had been his greatest predecessor. Aristotle made some gross misstatements in biology; but if one ignores these, as if the pages with them were missing, what is left is most impressive. Also, Aristotle knew a great deal about sex, in man and beast, but he always spoke of it with professorial decorum; he did not seek, and find, a pederastic sex symbol in every corner of Greek individuality, as Plato morbidly did.

Furthermore, Aristotle founded the systematic academic disciplines of Politics, Ethics, Rhetoric, and Literary Analysis.

Furthermore, he was the first to compose, in his *Physica,* systematic essays on Infinitude, Space, Void, and Time. He was the only one to transmit Zeno's ever fresh puzzles on motion. And, above all, only through Aristotle, and not in the least through Plato, is it known how much Pythagoreans contributed to the rise of Greek intellectuality.

Furthermore, Aristotle created, out of nothing—and he knew that it was out of nothing—the vast field of Logic.

Finally, the 19th century knew, although the 20th century tends to forget, that Aristotle created, out of little, the academic discipline of History of Philosophy. Every professional historian of philosophy is a descendant of Aristotle. And if he elects to be laudatory of Plato and derogatory of Aristotle, then he knows how to be so because Aristotle, and not Plato, taught him how to be so.

ATHENAEUS (c. A.D. 200), grammarian, composed a lengthy work pretending to be a report on dinner-table conversations among philosophers. It is a rambling, tedius work on an extraordinary variety of topics which might conceivably be conversation topics; the work is of importance as a source of many names, fragments of quotations, and bits of information.

BACON, FRANCIS (1561–1626), greatly admired by some as a "new-breed" scientific empiricist; somewhat boastful and charlatanish; and apparently not a paragon of honesty in public life.

On an occasion, with reference to himself, Bacon defines a philosopher "as having a mind sufficiently mobile for recognizing (what is most of all) the similarity of things, and sufficiently fixed and intent for observing the subtleties of differences." (*Collected Works of Francis Bacon,* new edition by Basil Montague, Vol. XV, 1834, p. 220.) That is, a philosopher is somebody seeking one-ness in many-ness, and discerning diverseness in apparent sameness. I think that this view of the role of a philosopher is, in effect, due to Plato.

The corresponding requirement for mathematics is this, that various types of "equalities," "equivalences," "congruences," "homeomorphisms," etc. between objects of mathematics must be discerned, and strictly adhered to. However this is not enough. In mathematics there is the second requirement that one must also know how to "operate" with mathematical objects, that is, how to produce new objects out of given ones. Plato knew philosophically about the first requirement for mathematics, but not at all about the second. And Greek mathematics itself never developed the technical side of the second requirement, and this was its undoing.

It is true though that, creatively, for the successful pursuit of mathematics, the faculty of conceiving suitable hierarchies of "sameness" among objects of mathematics is inseparable from the faculty of "operating" with such objects, even if the latter faculty is the more pronouncedly mathematical one. Also, from some psychologically oriented approaches, the second faculty is even subordinable to the first one, in a sense. In fact, one could say, and it probably has been said, that mathematical "operations" follow certain patterns, and that the habituation for the recognition of such patterns is not different from the habituation for the recognition of patterns of "sameness"; and any assertion that mathematics is learnable by "all" would sooner or later have to take recourse to an argument of this kind, I think.

On the other hand, since the beginning of the 20th century, "analytical" philosophy has been developing some "operational" attributes of its own; and this philosophy is therefore sometimes wondering what the differences between itself and mathematics (or "theoretical" science) really are.

BACON, ROGER (1214–1294), English Schoolman, Franciscan; lifelong advocate of the importance of science and mathematics. He was always outspoken, sometimes even tactlessly so, took personal risks, and spent two lengthy terms of years in confinement. He was also very learned, and his knowledge was so wide that it even included Hebrew and Aramaic. It must nevertheless be said that he was not a profound thinker, in philosophy proper, or in any area of natural philosophy; and there is no memorable specific doctrine or saying of his to match his militancy and "progressiveness."

BAILLY, JEAN SYLVAIN (Paris 1736–id. 1793), astronomer and historian of astronomy; political orator. At the outbreak of the Revolution he entered public life, and after a stormy career he ended on the guillotine, November 12, 1793.

BARTHOLINUS, ERASMUS (Roskilde 1625–Copenhagen 1698), Dane, was first professor of mathematics, then of medicine, at the University of Copenhagen. In 1669 he discovered the phenomena of double refraction in Iceland spar (a variety of calcite).

BERNOULLI, JAMES (Basel 1654–id. 1705), Patriarch of the Bernoullis and, perhaps, the most original among them. Was one of the founding fathers of the Calculus of Variations and, above all, of Probability and Statistics.

BERNOULLI, JOHN (Basel 1667–id. 1748), younger brother of James, and father of Daniel Bernoulli, the hydrodynamist; he coined the name *brachistochrone*.

BOHR, NIELS (1885–1962), physicist of Copenhagen, creator of the atom of quantum theory. He was a master umpire of physics during much of his lifetime.

For anybody interested in the problem of what constitutes originality of achievement in science it is imperative to read in E. Whittaker's *A History of the Theories of Aether and Electricity,* II, 106ff., the account of the genesis of Bohr's model of the atom. Virtually every part of the model had been knowingly made by somebody else, and very recently too, and Bohr did no more than assemble the parts, in a sense. Physicists are well aware of this, but there is no debate at all among them as to whether Bohr's achievement was fully his.

BOLTZMANN, LUDWIG (1844–id. 1906), Austrian physicist, introduced probability into thermodynamics (1877) and found a probabilistic interpretation of entropy.

BOYLE, ROBERT (Lismore Castle 1627–London 1691), Anglo-Irish physicist and chemist was the 14th of 15 children, author of *The Sceptical Chymist* (1661). He speaks loosely and vaguely, and it is frequently difficult to say with what specific achievements he might be credited.

BROUWER, L. E. J., Dutch mathematician (born 1881) who also created, even before becoming reputed as a topologist, the philosophical method of Intuitionism for the logical foundations of mathematics.

BURCKHARDT, JACOB (Basle 1818–id. 1897). In addition to his ever popular work on the Italian Renaissance, there is a large-scale work of his on the civilization of the Greeks; but he in no wise thinks that the Greeks were wonderful, and classicists evade publicizing the work.

CANTOR, GEORG (Petersburg 1845–Halle 1918). He started out with a very special problem of Riemann, in very technical mathematics, but it somehow led him, step by step, to the theory of sets, and to immortality.

CANTOR, MORITZ BENEDIKT (Mannheim 1829–Heidelberg 1920). His *History of Mathematics* is one of those large-scale works by bearded gaslight-Victorians which the 20th century does not quite know how to supersede with whatever it might try to supersede them with.

CARDANO, GERONIMO (Tavia 1501–Rome 1576), Italian philosopher, mathematician, physician; he cheated Tartaglia out of the formula for the solution of an equation of the third degree, which goes under the name of Cardano's formula.

CARNOT, SADI (Paris 1796–id. 1832) created a new aspect of thermodynamics in 1824, in a pamphlet of which only

200 copies were printed. It was saved from falling into oblivion by E. Clapeyron, and also Lord Kelvin.

CARTAN, ÉLIE (Dolomieu 1869–Paris 1951), leading French mathematician with a large following. His most individual achievement was the creation of so-called symmetric spaces.

CASSIRER, ERNST (Breslau 1874–Princeton, New Jersey, 1945), was a neo-Kantian; he was not only trying to accommodate the Kant of the 18th century to the needs of the 20th century, but he was even trying to persuade himself that, in essentials, Kant had been anticipating the 20th century anyway.

CAUCHY, AUGUSTIN (Paris 1789–Sceaux 1857). His family were ardent legitimists, but they somehow survived the Revolution. When the Bourbons were displaced in 1830, Cauchy expatriated himself, but he returned on his own in 1838. However, neither then, nor in 1848, nor in 1852 did he take an oath of allegiance, but from 1854 on he was allowed to hold a professorship of astronomy without taking one.

Cauchy is generally known for his work in mathematics, mainly in analysis; however he also played a significant part in the erection of a theory of mechanics for continuous media, and he was also working successfully in optics.

CAYLEY, ARTHUR (Richmond 1821–Cambridge 1895), English algebraist; in the 19th century one of few leading English mathematicians who were not physicists.

CHÂTELET, ÉMILIE DE BRETEUIL, MARQUISE DU (Paris 1706–Lunéville 1749), formidable bluestocking. Among other things, she translated—perhaps with some assistance from Clairaut—Newton's *Principia* into French; in fact, she was "fort occupée à mon Newton" apparently to her last.

Although married, she kept house with Voltaire from 1733 on. But, toward her end she was having an additional "petite distraction" with the poet Saint-Lambert, captain of the Guards, 11 years her junior; and she died of childbirth. Voltaire would not be consoled. Émilie was no more; and the likeness in the bezel of the ring on her dead finger was that of the brave captain.

CHRYSIPPUS OF SOLI (280–204 B.C.). Died as the head of the Stoic School at Athens. His list of publications comprised 705 titles; they are all lost, perhaps because their style was abominable. He may have been the author of the intriguing notion of τό λεκτόν (the spoken, the uttered).

CLAPEYRON, EMILE (Paris 1799–id. 1864), French railroad engineer and physicist. His renown derives from a "commentary" on the work of Sadi Carnot, and without this commentary both he and Sadi Carnot might have been forgotten.

CLAUSIUS, RUDOLF EMANUEL (Köslin, Pomerania, 1822–Bonn 1888), German physicist; creator of the thermodynamic conception of *entropy,* which is dominant in present-day speculations in cosmology.

COMMANDINO, FEDERICO (Urbino 1509–id. 1575), mathematician, mostly known for his translations of Greek classics, but he also wrote some works of his own. He belongs to the late Renaissance rather than to an early phase of the Scientific Revolution.

COPERNICUS, NICOLAS (Torn 1473–Frauenburg 1543). Whatever his stature as astronomer and natural philosopher may have been, mathematically there was nothing "revolutionary" in his work. Mathematically, the first revolutionary astronomer was Kepler.

CORNFORD, FRANCIS MACDONALD (Eastbourne 1874–Cambridge 1943), English classical scholar; wrote important commentaries to several dialogues of Plato, and a major commentary to Aristotle's *Physica,* in the Loeb Collection.

COULOMB, CHARLES DE (Angoulême 1736–Paris 1806). The so-called Coulomb's law in electricity (1789) was the first quantitative law in electrodynamics, and it is a puzzle why the first such law came so late.

COURNOT, ANTOINE AUGUSTIN (Gray 1801–Paris 1877), introduced mathematical formulation into economics in 1838. In other social sciences mathematization came much later, and this seems to give to economics a certain over-all advantage.

D'ALEMBERT, JEAN LE ROND (Paris 1717–id. 1783), was a star in the galaxy of mathematicians of the 18th century who were advancing mechanics. He is credited with having initiated the "principle of d'Alembert," which mathematically "transforms" a dynamic system of particles into a static one, and views equations of motion as if they were conditions for equilibrium; he initiated the erection of the wave equation; and he anticipated the Cauchy–Riemann equations of conjugate harmonic functions.

DEMOCRITUS OF ABDERA (460–350? B.C.), the younger partner of the founder of atomism Leucippus, was the "prince" ($\mathring{\alpha}\rho\iota\sigma\tau o\varsigma$) of Greek physics. In his approach, corpuscles are beginning to be real "particles"; that is, they are beginning to have properties of their own, presumed ones, which are not directly manifest in the sense-perceived macroscopic matter of which they are the constituents.

Plato does not mention Democritus, just as he does not mention Anaximander and Anaximenes; and in the case of Democritus this was interpreted to mean that Plato had been jealous of him (Diogenes Laertius IX, 40; Loeb Edition, II, 449–451).

Aristotle mentions Democritus frequently and affirmatively, even if he does not accept atomism for himself. Atomism necessarily imparted a certain measure of reality to the Void, which Aristotle somehow "abhors." However, Aristotle evinces a remarkable understanding for the tenet of the atomists that if there is to be a detailed theory of atoms, which are "full," there must be a detailed corresponding theory of the void (= space = place), which is "empty." If translated into present-day physics, this tenet avows that one cannot have "particles" without dual "fields"; or, more radically, that one cannot have "corpuscles" without dual (de Broglie) "waves."

Archimedes, in his *Method,* mentions Democritus as somebody likely to suggest a mathematical theorem, without necessarily being able to prove it; which is typical of physicists, in a sense.

The "ethics" of Democritus is typically that of a physicist who, secure in his standing as a scientist, feels impelled to discourse on "civics" too. It is fundamentally the ethics of a "conservative"; but there is a dash of "libertinism" added, just to show that one is "broad-minded" too.

In atomism, after Democritus came Epicurus (341–270 B.C.). Then, after more than 2,100 years, in A.D. 1841, Karl Heinrich Marx (1818–1883) began in earnest his career of social analyst with a lengthy doctoral dissertation, made up of dreary profundities, under the title *Differenz der demokritischen und epikureischen Naturphilosophie,* and, in the process, he affiliated the nascent social doctrine of his with a specific approach to professional physics and science.

De Morgan, Augustus (Madras 1806–London 1871), mathematician and logician, was one of several 19th-century British mathematicians of consequence whose knowledge of "higher" mathematics was apparently quite circumscribed.

Diogenes Laertius, "probably" lived in the 3rd century A.D. He wrote a surviving book on the *Lives of Philosophers* which is indispensable. But it has palpable limitations, and every classicist finds himself duty bound to denounce it, some time or another.

Diophantus, arithmetician, lived for a time in Alexandria, perhaps around A.D. 250. The "inward" style of his mathematics seems to be radically different from that of the "geometers" like Archimedes, and it appears to be impossible to find out where this different style came from.

A Greek edition with a Latin translation of his surviving work was made by G. Bachet in 1585. Fermat owned a copy of Bachet, and he apparently used it as bedside reading.

Dirichlet, Gustav Lejeune (Düren 1805–Göttingen 1859), German mathematician. He was a number theorist at heart. But, while studying in Paris, being a very likeable person, he was befriended by Fourier and other like-minded mathematicians, and he learned analysis from them. Thus equipped, he was able to lay the foundation for the application of Fourier analysis to (analytic) theory of numbers.

Du Cange, Charles Du Fresne, Seigneur (Amiens 1610–Paris 1688), a leading French erudite of the 17th century. Best known are his two glossaries, one for medieval Latin and one for medieval Greek. Regrettably, they are not very serviceable for scientifically oriented knowledge.

DUHEM, PIERRE (Paris 1861–Cabrespine, Ande 1916), physicist and leading historian of science. His academic career was most uneventful. After finishing the *École normale supérieure* in 1882, he was *maître de conférence,* first in Lille then in Rennes, and, from 1895, he was professor at Bordeaux, always in the *faculté des sciences.*

In addition to his many volumes in history of science, Duhem also published books and treatises in physics and chemistry, two volumes on *Energetics* among these; and, apparently, he wanted to be taken (and esteemed) as a physicist rather than as a historian.

EINSTEIN, ALBERT (Ulm, Germany, 1879–Princeton, New Jersey, 1955), was the most representative physicist of the first half of the 20th century and, by popular acclaim, one of the greatest ever.

To the general public, he was the "mathematician" who created the metaphysical problem of the meaning of simultaneity for events that are spatially apart, and he was also the creator of the equivalence between energy and mass, both venerable concepts, $E = mc^2$. To the physicist, he is furthermore one of three Founding Fathers of quantum physics, the other two being: Max Planck before him, and Niels Bohr, younger than himself.

In the mid-1920's there was a Young Physicists' revolution in quantum physics, which changed the face of physics, and of the earth, and of the universe. And the three Founding Fathers reacted differently to it.

Niels Bohr instantly made common cause with the revolutionaries, wholly, wholeheartedly, and unstintingly.

And he was rewarded with a position of eminent leadership of the movement, which he held for life.

Planck carried on as usual, and said even less than usual; but in a *Scientific Autobiography,* which he composed quite late, in 1948, he gave vent to some stored-up resentments, which I would not have expected him to keep so fresh for so long. But I had known him by his classroom manner only.

Einstein grew ever more into a critic of the novel theory, and he refused to legitimize some of its tenets to the last. His minority view was not wholly without merit, and he had a following, albeit a small one, even among quantum physicists themselves. But, for instance, the senior quantum physicist Max Born, who had lavished admiration on Einstein from his earliest, was distressed over Einstein's negative attitude and called it a "tragedy" (Paul Arthur Schilpp, *Albert Einstein: Philosopher-Scientist,* the Library of Living Philosophers, Vol. VII, 1949, pp. 163–164).

It is a remarkable phenomenon that most affirmers of Einstein cannot insist enough on his exploits in Space-Time, but pay scant heed to his achievements in physics of matter. Yet, in 1905, Einstein transformed Planck's radiation hypothesis into a postulate of quantization of all light; which postulate has been corroborated in numerous ways, much more convincingly than any of Einstein's contentions in General Relativity. And, in 1917, Einstein articulated the postulate of occurrence of stimulated emission of radiation; and Masers and Lasers, for instance, are an everyday corroboration of this.

There probably is no single-sentence explanation of this phenomenon; but if there were one, it ought to be the following. Einstein's achievements in physics of matter pertain to quantum theory, and to the age before the promulgation of the Uncertainty Principle; by spectacularity, however, anything in quantum theory before the advent of this Principle is a background event only. Just as in Special Relativity, anything before Einstein's cogitation on the meaning of simultaneity of space-separated occurrences, is there, by spectacularity, a mere background event too.

On the other hand, some physicists—not many, but the physicist-historian E. T. Whittaker (1873–1956) among them—have maintained that, spectacularities apart, the Special Theory of Relativity had been created not by Einstein but by Poincaré, Lorentz, and others, before him. For a balanced account see Charles Scribner, Jr., "Henri Poincaré and the Principle of Relativity," *American Journal of Physics,* 32 (1964), pp. 672–678.

EMPEDOCLES OF AGRIGENTUM (495–435 B.C.), philosopher, physician, evolutionist; genius, charlatan; publicity seeker, the only Western philosopher who deified himself. Born into high aristocracy, he disdained political power, preferring the "power over men's minds."

Aristotle accepted from Empedocles the doctrine, although in a modified setting, that there are four elements or, as Empedocles termed them, "roots": earth, water, air, fire; but Aristotle did not know how to rationally interpret the assertion that the vicissitudes of the universe are governed by (Empedoclean) forces of Love and Strife. Plato however liked these forces of Empedocles and cloaked them in sex symbolism (*Sophist* 242 C-E).

Empedocles wrote in poetry, in a style which, being both turgid and "Romantic," is inebriating to the susceptible. In his *Poetics,* Aristotle mentions, disapprovingly, that Empedocles wrote about philosophy in Homeric hexameter. But he chooses to forget, or it somehow does not come to his mind, that Parmenides had already done so before.

EUDEMUS OF RHODES, pupil of Aristotle, reputed author of histories of geometry, of arithmetic, and of astronomy.

When late commentators assert that they are quoting from Eudemus, it is never certain how literal the quotation really is.

EUDOXUS OF CNIDOS (408–355 B.C.), astronomer, mathematician, geographer; philosopher, orator, even office-holder.

When a "student" in Plato's Academy, Eudoxus lived in the Piraeus, where apparently the rents were cheaper, and he laboriously "commuted" to Athens on foot. After his "graduation" from the Academy, his relation to Plato was an ambivalent one, sometimes friendly, sometimes tense.

Not a line of Eudoxus survives. But Aristotle, in Book XII of the *Metaphysica,* prominently credits him with the construction of a planetary system which was somehow based on concentric spheres. And Archimedes, in an astronomical context in the *Sand-Reckoner,* somehow injects the name of Eudoxus. And this is in the context the only name which he adduces before making a filial sounding mention of his own father.

Furthermore, in the preface to Book I of his treatise *On the Sphere and Cylinder* Archimedes credits Eudoxus with the proof of the theorems that the volume of a pyramid is one-third of the volume of the prism which has the same base and equal height, and that the volume of a cone is one-third that of the cylinder with the same base and equal height. And the manner in which Archimedes speaks of Eudoxus suggests that, in the judgment of Archimedes (if he had to render one), Eudoxus was the greatest (Greek) mathematician before him—which he may well have been.

Secondary documentation suggests that Eudoxus was a specialist in the theory of proportions as expounded in Book 5 of Euclid, but there is no documentation for an assertion that Euclid actually copied Book 5 from, or even closely modeled it on, a treatise of Eudoxus.

That Eudoxus also had bona fide interests in philosophy is taken for granted by Aristotle. In Books I and XIII of the *Metaphysica* he adduces some views of Eudoxus' relative to Plato's theory of Ideas; and in Book X of the Nichomachean Ethics, Aristotle reports that Eudoxus had a hedonistic but not at all libertinistic outlook on morals.

Other competences of Eudoxus are reported by sources later than Aristotle and Archimedes. He was such an imposing name in later antiquity that "Eudoxus" invariably stood for "Eudoxus of Cnidos."

Somebody ought to write a readable book about *The Life and Times of Eudoxus*. The article about Eudoxus in Pauly-Wissowa, *Realencyklopädie,* by F. Hultsch, although admirably exhaustive, is even more "unreadable" than articles in Pauly-Wissowa sometimes are.

EULER, LEONHARD (Basel 1707–Saint Petersburg 1783), Swiss mathematician. He went to Russia because he had a large family to provide for, and Catherine the Great paid him enough. During a good part of his life his eyesight was deteriorating; and for a number of years he was totally blind and working through assistants.

His mathematical output was staggering. He certainly was the most productive mathematician ever and, in the view of some, the greatest. His name occurs in virtually every area or subarea of mathematics which was not removed from his times. He also contributed greatly to mechanics. His theory of rigid bodies is a model of a theory in mechanics, and his greatest single achievement. Euler also wrote many books, all of which are interesting to read; his *Algebra,* for instance, is diverting and even charming.

Euler was not a stirring philosopher, but he showed remarkably good sense when divining subsequent developments in physics proper. He was the only savant of the mid-18th century who did not regard heat as a material substance; he asserted that electricity and light are propagated by the same aether; and he expressed the opinion that the phenomenon of light pressure is just as likely to occur on the undulatory hypothesis as on the corpuscular one. (See E. Whittaker, *A History of the Theories of Aether and Electricity*, I, 40, 98, 274.) We are not excerpting the pertinent passages from Euler, because the contexts in which they appear, as a whole, are somehow more ingenuous than the intentionally excerpted passages might suggest.

EURIPIDES (c. 480–406 B.C.), Athenian, the last of the three Greek tragedians, the other two being Aeschylus and Sophocles. Euripides, finally, was a realist; all characters of his are "human," if rarely humane, even in the *Bacchae,* which, incidentally, is the only source work for the knowledge of the Dionysian cult. Renowned for its timeless realism is the scene, in Corinth, between Jason and Medea—between the "civilized" Greek and the "barbarian" first-generation immigrant.

Some social strata of his generation and of several generations to follow were hostile to Euripides for finding themselves mirrored, faithfully, in his realistic "bourgeois" portraits. The comedy writer Aristophanes, a conservative in his attitude towards traditional civic morality, deemed him a "subversive"; the name of Euripides was meant to be sullied by the rumor that his mother had tended a greengrocer's stall, and by rumors of his having died in some unusual manner; and a ladies' club once opened a drive to have him taken off the repertory stages for being unfair to the female sex in his characterizations.

However, as many as 18 plays of his have survived, because, in fact, he was popular with directors, actors, and audiences; they could understand him without special briefings in Greek mythology and history, although both enter heavily into his plays.

EUTOCIUS (6th century A.D.), wrote commentaries on Apollonius and Archimedes. He must have been antiquity's last commentator on mathematics.

FECHNER, GUSTAV THEODOR (Gross-Särchen 1801–Leipzig 1887), philosopher, and also physician and physicist; co-author of the Weber–Fechner law that "sensation varies as the logarithm of the stimulus."

This law has apparently not initiated a trend toward a genuine mathematization of psychology. Whatever other mathematics there is in psychology comes to it as a part of the statistics which it applies, but not out of its own indigenous resources of conceptualization.

FERMAT, PIERRE DE (Beaumont-de-Lomagne 1601–Castres 1665), from 1631 on counsellor of the (local) parliament of Toulouse by profession; and mathematician by passion. He was a great analyst, with a *penchant* for number theory and admiration for Diophantus. In a leading memoir on maxima and minima, Fermat even tries to convince himself that maxima and minima were already studied analytically by Diophantus; which is absurd.

FOURIER, BARON JOSEPH (Auxerre 1768–Paris 1830), French mathematician, and mathematical physicist, author of a standard work on the propagation of heat, in which he introduced, for all time to come, the so-called Fourier series and Fourier integrals.

He accompanied Napoleon to Egypt and was French commissioner to the Sultan; afterwards, as prefect of Isère, he built the road from Grenoble to Turin via Mont-Genèvre.

FRESNEL, AUGUSTIN (Chambrais (Broglie), Normandie, 1788–Ville d'Avray 1827), great French physicist, the true founder of mathematical optics in its details. He was the son of an architect, of delicate health, and slow of learning in his early youth. By profession he was an engineer, and a hardworking one; and his earnest research work in optics began in 1814, when, due to anti-Bona-partist sentiment, he was temporarily out of occupation. He died of consumption at the age of 39; and in his collected works, which fill three bulky volumes, many an important memoir is taken from an inedited manuscript.

Fresnel established the principal laws of optics, such as those for reflection, refraction, total reflection, diffraction, aberration, etc. "Popular" accounts of him do not know what to make of these theoretical achievements, and they cling to his one "spectacular" accomplishment of introduc-ing compound lenses instead of mirrors for use in light-houses. "Serious" historical studies of Fresnel also have their difficulty. Today, Fresnel's eminence resides in the fact that his laws of optics have, long after him, bravely withstood, more or less, profound changes of outlook on the basic nature of light and its mode of radiation; but, for a student of history to duly emphasize this would violate the schoolmasterly mandate that the course of past events must be presented only "wie es eigentlich gewesen."

GALILEO, GALILEI (Pisa 1564–Arcetri 1642) was, by textbook consensus, the Leader of the Scientific Revolu-tion. In my personal judgment, the Scientific Revolution was too elemental and unparalleled an occurrence to have had a textbook-type leader. But if one insists on "leaders" for this revolution too, I would opt for Kepler as its actual Originator, even if he was somewhat younger than Galileo, for Newton as its Chief Executor, and for Galileo as its Most Articulate Popularizer.

Galileo's specific achievements were: (i) the discovery of the Jupiter satellites and, then, of sunspots, and altogether his insistence on the systematic use of the telescope; (ii) his formulation of the law of the falling body, and his discovery of the parabolic trajectory in vacuum; (iii) his initiation of thermoscopy; and (iv) his reflections on the elasticity properties of a beam, even if in this area his findings were nothing but "discorsi," that is, conversationally qualitative suggestions only.

In one respect Galileo never aged; the virulence of his anti-Aristotelism never abated.

GAUSS, CARL FRIEDRICH (Brunswick 1777–Göttingen 1855), THE mathematician according to some; astronomer; physicist.

As if to inaugurate the 19th century, his *Disquisitiones Arithmeticae* appeared in 1801. But the 19th century in mathematics arrived in earnest somewhat later, when Gauss saw himself "motivated" to produce four different proofs for the assertion that a polynomial with complex coefficients has a complex root. It is reported that the ageing Lagrange was puzzled by this attitude; from his 18th-century approach it was indicated to have just one proof, a satisfactory one, and to continue from there with something else.

GAY-LUSSAC, LOUIS JOSEPH (Saint-Léonard-de-Noblat, Marche, 1778–Paris 1850), leading chemist, author of the law named after him. In 1804 he also made "daring" balloon ascensions, in which he made observations on magnetism and on temperature and humidity and collected several samples of air at different heights.

GEMINUS (1st century B.C., probably), probably a pupil of Posidonius; wrote apparently some kind of encyclopedia of mathematics, with emphasis on its history and its philosophy and on classification of mathematical disciplines. To judge by the remnants traceable to it, its loss was not painful to mathematics itself, although of course regrettable in the history of mathematics.

GREEN, GEORGE (Sneinton, near Nottingham, 1793–id. 1841), businessman, dilettante mathematician, creator of the word "potential," and of "Green's formula."

GROSSETESTE, ROBERT (Stradbroke in Suffolk, around 1170–Lincoln 1253), leading Oxford don; teacher to Roger Bacon, who admired him, and lecturer to Franciscans; bishop of Lincoln from 1235; was present at Runnymede in 1215 when King John signed the *Magna Charta.*

Grosseteste inaugurated a kind of pre-modern science and mathematics of science, namely a medieval and Renaissance science, comprising calendar reform, optics (Grosseteste somehow knew about magnifying glasses), rainbows, meteorology, heat of the sun, kinematics, etc.; and this science more or less corresponded to the pre-modern mathematics of arithmetics and algebra, which was inaugurated by Robert's near-contemporary Leonardo da Pisa (Fibonacci). But I do not wish to suggest that there was any awareness of such a parallelism at the time, or that the fact of there being such a parallelism was a creative force in itself.

However, whereas the pre-modern mathematics produced, beginning with Fibonacci himself, several successive large-scale textbooks which led to such specific achievements as the solution of equations of the third and fourth degrees and the introduction of symbolism by Viète, the pre-modern science, which began with Grosseteste, did not culminate in comparable, theoretically oriented, specific prerequisites for the rise of modern science in the 17th century. As against this, some contemporary historians maintain that pre-modern science had created and made ready a certain methodology and theory of science, by which the rise of actual scientific activity in the 17th century was conditioned and guided. This, however, is a very speculative approach. It is difficult to discern a method of science other than within scientific activity proper at its fullest; and a philosophy of science can be reliably obtained only after scientific activity proper has been in progress.

Perhaps as a legacy of neo-Platonism, in Grosseteste, as in other pre-modern scientists, physical light was endowed with certain metaphysical attributes, which, apparently, could be "experienced" and even "communicated," but not "explained."

HARRISON, JANE ELLEN (Cottenham, Yorkshire, 1850–London 1928), a first-rate lady classicist who does not receive enough recognition because she was a woman. Her leading work is the *Prolegomena to the Study of Greek Religion* (1903). Every single statement in this work can, and probably has been, harpingly contested; but the book as a whole is an eye-opener.

HELMHOLTZ, HERMANN LUDWIG FERDINAND VON (Potsdam 1821–Charlottenburg 1894), an important name in 19th-century physics.

He studied medicine on a state scholarship which obligated him to accept state employment for a number of years. His paper on *The Conservation of Energy* (1847) was published while he was doing "enforced" service as an army physician; but on the strength of this paper he was released the following year from further such obligations although he still owed three years to the state.

Helmholtz lost many a prospective pupil by giving lectures at the period from one to three o'clock in the early afternoon, which was then the customary midday-dinner and siesta time. But Helmholtz did secure the pupilship of Heinrich Hertz, and a very devoted pupil he turned out to be.

HERACLITUS OF EPHESUS (535–475 B.C.) is immortalized by sayings that "all is in flux," that "one cannot step into the same river twice," and that opposites meet and become indistinguishable. These sayings are the man; and before reading books about these sayings, and about how great a natural philosopher Heraclitus was, it is well to consider that the sayings are self-commenting by their brevity, mystique, melancholy, and, above all, by their unresolvable equivocation.

The idiom of Heraclitus was poetry-in-prose of a novel kind, and this language must not be overinterpreted, lest the interpretation yield explicated rationalizations which Heraclitus probably did not invest into it. For instance, the Heraclitean saying (fragment 84a)

"Metaballon anapauetai (though it changes, it stands still)"

fares better when viewed as a self-sustaining aphorism than as a sentence from an alleged "system" of science. As an aphorism it is immediately and fully comprehensible as it stands, and superior to its 19th-century counterpart

"Plus ça change, plus c'est la même chose"

(January 1849, by Alphonse Karr 1808–1890) of vaudevillesque origin (see P. Duprée, *Encyclopédie des Citations,* Paris 1959, p. 111). But as a scientific statement the saying 84a of Heraclitus is "archaic," and semi-articulate, and in need of a commentary as to what was going on in science before and after.

Heraclitus was maladjusted; he suffered painful disappointments, and he wanted the world to stand still and become aware of this. And when the world refused to do so, he withdrew into misanthropic solitude, and died, in misery, of some lingering disease.

Such reports about Heraclitus, which come mainly from Diogenes Laertius, are perhaps, as frequently asserted, nothing but legendary backward reconstructions from known sayings of Heraclitus himself, which used to have wide currency in late antiquitity; but, even so, it is indisputable that the personal manner of Heraclitus was abrasive and uninviting. Thus, in his literal fragments Nos. 42 and 40 (Diels-Kranz) Heraclitus crudely denigrates the stature of Homer, Archilochus, Hesiod, Pythagoras, Xenophanes, and Hecataeus. But in fragment 39 he extolls the "sage" Bias of Priene, a person of no historical stature, whose only claim to fame is his motto—which must have endeared him to Heraclitus—that "most men are bad." Also, many of the known sayings of Heraclitus speak of dying or killing; or of ecological filth; or of man's intellectual somnolence and stupor. It must be said however that, nevertheless, the hearts of commentators go out to Heraclitus, and that even the staidest among them are anxious to iron out wrinkles in the character of the Old Curmudgeon of our philosophy. A bibliographic compilation of views about Heraclitus is given in the work of E. Zeller and R. Mondolfo, *La filosofia dei Greci nel suo sviluppo storico, Parte I, I Presocratici, Volume IV, Eraclito* (Firenze, 1961).

Heraclitus created the literary genre of the philosophically charged but intentionally obscure, that is un-explicated, aphorism, and his own name for this brand of aphorism was "Logos." When, after Plato and Aristotle, and after Alexander the Great, weariness began to descend upon Hellas, the Stoics began to turn, or re-turn to the Heraclitean Logos; and, after a small part of it became Stoic Logic, the major part of it became gradually cloaked in Mysticism and Eschatology.

Heraclitus has caused mischief by his sayings (fragments 80 and 13), "War-might is right," and "War is a king who makes slaves and freedmen, men and gods."

Plato was intrigued by Heraclitus, and Aristotle says that Plato was much influenced by him toward becoming the kind of philosopher that he turned out to be; yet Plato did not name a dialogue after Heraclitus, or make him an interlocutor in one.

Aristotle created the pungency of the saying, "All is flux"; the original version of the saying of Heraclitus is presumed to have been less compacted (see W. Capelle, *Die Vorsokratiker,* 1935, p. 132, n. 1; and some books about Heraclitus). But, as usual, Aristotle also asked himself what, if anything, Heraclitus contributed to the problem of the physical structure of matter. Heraclitus speaks a great deal about some kind of a "Fire," and at times Aristotle toys with the idea that this Fire might be a "substance." But at other times Aristotle thinks differently, and rightly so, and he ruminates whether it might not be some kind of "emanation," or even a kind of "soul-like" ectoplasm (*De Anima,* Book I, Chapter 2; 405 a 25). However he never makes a decision, because none can be made; today, there are many more views in addition to these, and there is no prospect of a consensus.

HERODOTUS OF HALICARNASSUS (480–425 B.C.), great sociologist and anthropologist; master teller of stories from the history and the background of the history of the relations between Greeks and Persians which led to the Persian Wars.

It must not be assumed that his competence as a "scientific" historian was affected by his religious belief that if a man takes his happiness for granted the gods become "jealous" and inflict unhappiness upon him.

HESIOD (8th century B.C.), an early Greek poet. He composed a *Theogony,* which somehow is viewed as being also an "early" cosmogony. And, beginning already with Plato and Aristotle, this somehow made him also into an "early" philosopher; although in the case of Plato this "elevation" of Hesiod to the status of a philosopher (in *Cratylus* 402 B C he is made into a pre-Heraclitean) may have been a pleasantry only.

In the path of development from Homer and Hesiod to Thales, just before reaching Thales, textbooks also deal with a mythographer Pherecydes of Syros, already mentioned in Aristotle; but not much is lost if one quickly turns the pages about him.

HILBERT, DAVID (Königsberg 1862–Göttingen 1943), German mathematician; did striking work in algebra, theory of numbers, functional analysis, axiomatics, and logic of mathematics.

HIPPASUS OF METAPONTUM (around 500 B.C., say) was an early member of the Pythagorean fraternity, and, apparently, he was interested in Pythagorean science, but not in its "mysticism." At any rate, late tradition has it, more or less, that he "betrayed" Pythagorean "secrets"— apparently scientific ones—to outsiders. This infuriated some fellow members; they accused him of venality, and wished him dead. Some even declared him "officially dead" and they erected a funeral column in his "memory."

HIPPOCRATES OF CHIOS (5th century B.C.), a leading mathematician (not to be confused with Hippocrates of Cos, the 5th-century B.C. physician).

The mathematician was snatched from the brink of near-oblivion at the last possible moment, namely by Simplicius in the first half of the 6th century A.D. Aristotle, in *Physica* I,2, makes reference to "the squaring of the circle by means of segments." Simplicius in his great commentary on the *Physica* suggests that Aristotle is thinking of Hippocrates, and he proceeds to quote a description of the work of Hippocrates from the *History of Geometry* by Eudemus, of which Simplicius feels fortunate enough to own a copy. Simplicius insists that he is going to adhere to the literal text of Eudemus, except, alas, for interweaving explanatory comments of his own based on Euclid; and during the past 100 years generations of our scholars have been trying to unweave the weft.

HUIZINGA, JOHAN (Groningen 1872–De Steeg 1945), a Dutch intellectual-historian of first rank. He even helped to unmask a (contemporary) copyist of Vermeer.

HUMBOLDT, ALEXANDER VON (Berlin 1769–id. 1859), world traveler, explorer, naturalist; universalist of knowledge about nature. He was something of a belated Romantic, and in Germany innumerably many peaks, rivers, and woods are named after him.

HUYGENS, CHRISTIAN (The Hague 1629–id. 1695) was the most outstanding physicist proper of the Scientific Revolution, and he was also a remarkably gifted mathematician. But he was not at all a "natural philosopher," either inspired or inspiring; there is not a single "memorable," or only "quotable," sentence or paragraph of his, and certainly no kind of "Scholium" with which to agree or disagree. He must have been a very lifeless person with whom to spend an evening, unless one let him play his musical instrument or instruments through the length of it.

JACOBI, CARL (Potsdam 1804–Berlin 1851), excelled in mathematics and mechanics. He introduced, in German universities, the giving of organized lecture courses in really advanced topics of mathematics. For instance, Jacobi himself gave very intensive lecture courses in elliptic functions.

JOACHIM, HAROLD HENRY (London 1868–Croyde, Devon 1938), philosopher and Greek scholar. He was a master interpreter of Aristotle, and his commentary on the treatise *De Generatione et Corruptione* is a major work on Aristotle as such.

KANT, IMMANUEL (Königsberg 1724–id. 1804) was the Deifier of Space and Time; some philosophers, and even scientists, are still under the spell of this deification.

Kant's *Critique of Pure Reason* (1781, 1787) is by publication date synchronous with the *Mécanique analytique* of Lagrange (1786), but by mathematics it is 2,000 years behind it. In the *Critique,* mathematics still consists of the geometry of Euclid, and of the ordinary arithmetic of integers. There is no awareness of the power of operation with symbols, as in Viète and Descartes, or of the power of the infinitesimal calculus and analysis in general; these powers had erected the edifice of Rational Mechanics by then.

Also, Kant's *Critique* "validates" mathematics by verifying that a statement like $7 + 5 = 12$ is a synthetic judgment a priori; and Kant even resorts to counting integers off on the fingers of his hand. But this is only "anthropological" mathematics, and not yet the "organized" mathematics which created Rationality through Literacy.

KEPLER, JOHANNES (Weil, Wurtemberg, 1571–Ratisbon 1630), a cool-headed, most "modern" rationalist, masquerading, for some kind of intellectual relief, as a still-medieval irrationalist and mystic; creator of modern physical astronomy; the first who dared to break out of the circle, be it simple, or epicyclical.

Without Kepler there would have been no Newton; and without Newton's "explication" of the three Keplerian laws, Kepler would not have lasted long.

KIRCHHOFF, GUSTAV ROBERT (Königsberg 1824–Berlin 1887), important German physicist.

"Old-fashioned" popular accounts of him, some even in very recent encyclopedias, make mention, first of all, of his law on the branching of a current in an electric network. In the second place they credit him with the founding, partly in collaboration with R. W. Bunsen, of systematic spectroscopy and with an explanation of Fraunhofer lines.

More modern accounts put the two achievements on an equal footing.

And really modern accounts heavily emphasize the second achievement; and they also state that in connection with this Kirchhoff introduced the conception of black body radiation or, what is the same, of cavity radiation, and established important theorems about emission and absorption of it.

It does not necessarily reflect on the veracity or meaningfulness of a finding in history of science if the finding presents itself in versions which depend on the time when made; the variability of the versions of the finding is then simply a part of the history of the event which is to be "found."

LAGRANGE, COMTE LOUIS DE (Turin 1736–Paris 1813), one of the greatest mathematicians ever. His hallmark is his *Méchanique analytique* (1786), the master textbook in its subject. Many results from this work, which are attributed to Lagrange as a matter of course, go back to achievements which were conceived and initiated, and frequently even fully accomplished, by Euler before him. But the "generalized coordinates" of our mechanics of today were conceived and installed by Lagrange, and this was an achievement of unmatchable magnitude.

From our distance, the space of generalized coordinates (or, what is the same, of free parameters) of Lagrange, if viewed as an entity in its own right, may be interpreted to have been an in-between creation in the transition from the traditional space of experiential and philosophical awareness, as still in Newton and Kant, to purely mathematical space as manifesting itself in various geometries of the 19th century (non-Euclidean, projective, differential, and algebraic geometry; topology of manifolds). But this "transitional" space also had "direct" descendents of its own, such as the spaces underlying "equations of state" in thermodynamics, and also "phase spaces," as in the work of Boltzmann and Gibbs. Even the spaces of quantum theory and quantum field theory, although "purely mathematical" by intentional construction, have retained an aspect of descendency direct from Lagrange.

LAPLACE, MARQUIS PIERRE SIMON DE (Beaumont-en-Ange, Normandie, 1749–Paris 1827), mastered the mathematics, astronomy, physics, and theory of probability of his time, and contributed to all. He started out, from modest social origins, as a protégé of D'Alembert. Notable achievements of his are: a cosmological hypothesis; revision of Newton's formula for the propagation of sound; a memoir with Lavoisier on quantitative aspects of chemistry; and the introduction of "Laplace integrals" as "generating functions" into the theory of probability.

LEBESGUE, HENRI (Beauvais 1875–Paris 1941), mathematician; he could write engagingly simply, with a minimum of formulas and cross-references, about substantive matters of his creation, which are among the building blocks of 20th-century mathematics.

LEIBNIZ, GOTTFRIED WILHELM (Leipzig 1646–Hannover 1716). Whatever his greatness and importance may have been, it is a fact that, from whatever extenuating reasons, he never produced a single outstanding work or accumulated a single outstanding achievement, as did Newton in his *Principia*.

LEONARDA DA PISA, also FIBONACCI (Pisa 1175–after 1240), a "mystery man" of the history of mathematics. It is not known what "motivated" this son of a "minor vice-president" of an outlying branch house of a Pisa firm to become a professional mathematician rather than to continue in business too. Fibonacci broke a lull of many centuries in European mathematics by publishing a very skillfully assembled collective work, the *Liber abaci,* which brought him great fame, and perhaps even some fortune. Among other things, the work introduced the so-called Fibonacci numbers, and it familiarized the West with the Indo-Arabic positional system. But it took centuries before the use of the system became widespread.

LEVI-CIVITA, TULLIO (Padua 1873–Rome 1941), mathematician, many-sided, introduced the concept of parallel displacement in differential geometry. He had a first-rate teacher, namely G. C. Ricci.

LIE, SOPHUS (Nordfordeid 1848–Christiana 1899), leading mathematician, working, seemingly "narrowly," in one single field of his choice and creation, and nothing besides.

LITTRÉ, ÉMILE (Paris 1801–id. 1881), was not only a giant lexicographer, and an outstanding Greek scholar, but also an active and publishing philosopher of positivist persuasion, and even a man of public affairs.

LORENTZ, HENDRIK ANTVON (Arnhem 1853–Haarlem 1928), Dutch physicist of stature, chief representative of the transition of physics from its Maxwellian state to 20th-century habitats.

LORIA, GINO (Mantua 1862–Genoa 1954), mathematician and historian of mathematics. He was reluctant to speculate on the causations of historical trends.

LUCRETIUS (98–55 B.C.) in full: Titus Lucretius Carus; great Latin poet. His *De Natura Rerum* is a poem on the place of man in the universe; and it is in particular a large-scale exposition of the doctrine of Epicurus, of which

Lucretius was an adherent. The didactic content does not at all mar the poetry of the work.

MacCullagh, James (1809–1847), a "minor" but interesting mathematician and physicist. He died by suicide.

Mach, Ernst (Turas, Moravia, 1838–Haar, near Munich, 1916), Austrian physicist. He composed three histories on topics in physics, namely, on mechanics, on optics, and on thermodynamics. It does not reflect on Mach that only the first has become widely known and that the third has not even been translated into English as yet.

Maimonides (Cordova 1135–Cairo 1204), Spanish Jew; physician; Talmud scholar; theologian-philosopher of Aristotelian orientation; one of the greatest figures in world-wide scholasticism.

As a physician he wrote 18 works on medical matters; a few survive, and they deal with physical and dietary hygiene. Maimonides was physician to the court in Cairo, and the duties were time-consuming.

Yet, after first producing a commentary on the *Mishna* (1158–1165), Maimonides found the time to compile, accurately although relying much on memory, a monumental systematization and codification of the entire *Talmud* (1170–1180), and this was his most prodigious scholarly feat.

However, to scholars at large, Maimonides is known mainly for his *Guide of the Perplexed,* which is an Aristotle-inspired, large-scale attempt to harmonize Faith and Reason, and which exerted considerable influence on scholasticism everywhere. For instance, the proofs for the existence of God in Aquinas are all blueprinted in Maimonides. The *Guide* was written in Arabic, which was the leading idiom of philosophy at the time.

An important occurrence in the *Guide* is the report, in

Chapter 73 of Part I, on a version of extreme atomism of the Islamic Mutakallemim. In this atomism, everything is made discrete, that is "quantized": space, time, matter, motion. In motion, a corpuscle "jumps" from one space position into a neighboring one. The time unit required for the jump is always the same, and the "illusion" of faster or slower motion arises by the number of time units of "rest" which elapse between jumps. Only "molecules," that is, assemblages of two (or more) atoms, have a non-zero size; however, a single atom is a "point," and has dimensions zero. If macroscopic phenomena conflict with microscopic explanations, then this is due to the unreliability of our senses.

This atomism was developed for the sake of a strict Islamic orthodoxy, but it sounds much more "modern" than the atomisms of "freethinkers" like Democritus and Epicurus.

MARINUS (second half of the 5th century A.D.), neo-Platonist, biographer and successor of Proclus; wrote, among other things, a commentary on the *Data* which is ascribed to Euclid.

Marinus was much more sober in his thinking than the "typical" neo-Platonist, and it is not easy to decide whether he was so from a superiority over his environment, or from a lack of imagination, as *apparently* his detractors maintained.

MARIOTTE, EDME (1620–Paris 1684), French aerodynamist; leading discoverer of the so-called Boyle–Mariotte law; wrote also a book on logic. He was capable of clear and assertive enunciations, much more so than Boyle.

MAXWELL, JAMES CLERK (Edinburg 1831–Cambridge 1879), a Scot, whose name has become a synonym for 19th-century physics.

He created the field theory of Electromagnetism, and of Light, which, according to him, "consists in the transverse undulations of the same medium which is the cause of electric and magnetic phenomena." His self-chosen teacher, Faraday, had presented him with the Dielectric; and, as if from courtesy, Maxwell devised the Displacement Current to put into it.

In addition to being a field-theorist, Maxwell also knew a thing or two about large assemblages of similar minuscular objects. Just as Newton, in addition to his Gravitation, had his Optics, so also Maxwell, in addition to his Electrodynamics, had his Gas Kinetics, to which he contributed the "Maxwell distribution," in the main.

Maxwell's stature can be gauged by the fact that his comfortably written *Treatise on Electricity and Magnetism* is serviceable as a student's text in "classical" electrodynamics even in today's age of quantum field theory. The treatise is occasionally even compared to Newton's *Principia;* although, in Maxwell's *Collected Works,* which are dated 1890, the editor, W. D. Niven, towards the end of his Preface, wonders doubtingly what the future of the *Treatise* could be. But then, the Preface is not aware of the great confirmatory work of Heinrich Hertz, which appeared in 1889.

It is said that Maxwell's wife was a shrew; but he nonetheless achieved a large amount of work before dying, much too soon, at the age of only 48.

MELISSUS OF SAMOS (5th century B.C.), gentleman-philosopher; also led the Samian fleet to victory over the Athenians in 442 B.C.

Melissus was younger than Parmenides, and a partisan of Eleatism, albeit, apparently, from a geographic distance. He was an indifferent philosopher, and it is neither easy nor urgent to recall what it was that he taught; however, a strange circumstance secured for him a degree of "immortality" which exceeds his desert. Melissus imparted infinitude to certain attributes of the Parmenidean universe, which in Parmenides himself had been finite. This would put Aristotle into a rage whenever his thoughts would turn to Melissus, because for Aristotle the belief in the finiteness of the universe was a sacrament, and the only redeeming feature about the doctrine of Parmenides; and Aristotle, after having persuaded himself that Melissus had committed a deadly error in his reasoning, would denounce him as a boor and simpleton. This denunciation, however, conferred on Melissus the permanent status of someone insensibly maligned; and, ever since, philosophers in every generation have labored valiantly to uncover greatness in Melissus.

MOMMSEN, THEODOR (Garding, Schleswig, 1817–Charlottenburg 1903) is the author of important works on Latin Inscriptions and Roman Law, and of a much read *History of Rome*. In his *History* he acts like a Roman "patriot." He trembles at the thought that this "dreadful" Hannibal might have conquered Rome, and he feels eternally grateful to the Scipios—whom he identifies with the Prussian "Liberals" of his time and liking—for having saved her.

Other historians are very reluctant to judge adversely Mommsen's biases and the absence of Volume 4 of his *History*. For references see Bruno Weil, *2000 Jahre Cicero* (Zurich, 1962), especially pp. 324ff.

MONTUCLA, JEAN-ETIENNE (Lyon 1725–Versailles 1799), author of the great *Histoire des mathematiques* (1758)—note the plural "des mathematiques"—was also a co-founder of the *Gazette de France*.

NEWTON, SIR ISAAC (Woolsthorpe, Lincoln, 1642–Kensington, Middlesex, 1727), was—himself. In the realm of knowledge there has been no other book like his *Principia*. It was rooted, as mathematics, in a distant past, and it stretches, as physics, into a distant future. It will never age, because it was never young, and will always remain itself.

Newton's streamlinedness—amid quaintness—comes through, with discrete force, in his pronouncement on man-made satellites, which he offers in the prelude of the *Principia,* within Definition V, in the following words.

If a leaden ball, projected from the top of a mountain by the force of gunpowder, with a given velocity, and in a direction parallel to the horizon, is carried in a curved line to the distance of two miles before it falls to the ground; the same, if the resistance of the air were taken away, with a double or decuple velocity, would fly twice or ten times as far. And by increasing the velocity, we may at pleasure increase the distance to which it might be projected, and diminish the curvature of the line which it might describe, till at last it should fall at a distance of 10, 30, or 90 degrees, or even might go quite round the whole earth before it falls; or lastly, so that it might never fall to the earth, but go forward into the celestial spaces, and proceed in its motion *in infinitum*.

After Newton, in physics proper—that is, outside of mechanics—developments which eventually continued where he had left off were quite slow in unfolding; and we will adduce several instances of this.

Newton was wont to speak of electricity and magnetism as if expecting that others ought to do something about establishing laws for them. But the first specific statement about them that would really have met Newton's expectation, namely the Coulomb law, became known only in

1785, that is about 100 years after the first edition of the *Principia*.

Newton's discovery and analysis of the continuous spectrum of the sun had a worldwide appeal, and stirred everybody's imagination; and a concern of Newton's with thermic properties of the spectrum is implicit in some of the Queries. However, attempts to observe the heat intensity along Newton's spectrum are recorded only from the last decades of the 18th century (M. Landriani 1776, Abbé Rochon 1783, J. Senebrier 1785). And only in 1800 did W. Herschel venture out with a thermometer into the infra-red; and soon afterwards in 1801, J. W. Ritter with a simple chemical device into the ultra-violet.

Newton did not know about dark lines in the solar spectrum. And it was over 100 years before they were discerned by W. H. Wollaston in 1802 and, much more satisfactorily, by J. Fraunhofer in 1814.

Newton did not introduce a notion of energy; and it took over 150 years before a satisfactory conception of energy was arrived at in the 1840's.

In Query 6 of the *Opticks,* Newton virtually introduced, by name, a Black Body of thermal radiation, or a "Gray" Body at any rate. But only beginning in 1859 did G. Kirchhoff introduce a sharp definition of it, and promulgate the first incisive laws about it. And the spectroscopic penetration of sublunary and celestial matter could begin only then.

Lastly, I wish to point out that Newton's law of motion, "The change of motion is proportional to the motive force," does not correspond to the "acceleration equation"

$$m \frac{d^2x}{dt^2} = F$$

but to the "momentum equation"

$$\frac{d}{dt}(mv) = F$$

[347]

in which $mv \equiv m\ dx/dt$ is Newton's "quantity of motion" (= momentum); and that in the 20th century it was the approach to mechanics via the momentum equation which created first the theory of relativity and then the theory of wave mechanics.

NICHOLAS OF CUSA, also CUSANUS (Kues 1401–Todi 1464), German theologian, Cardinal; anticipated views of the early stages of the Scientific Revolution after 1600. The 15th century had very few really prominent "intellectuals," and "Cusanus" was one of the few.

NIEBUHR, BARTHOLD GEORG (Copenhagen 1776–Bonn 1831), German historian. He pioneered in sifting the facts out of the traditions about the story of Rome under the kings, and thereby became a teacher to historians.

By first occupation he was a finance expert, and he held high-ranking posts as such.

OCKHAM, WILLIAM OF (Ockham, Surrey, 1295–Munich, Bavaria, 1349), English Franciscan. He clashed with the Papacy and was excommunicated; and he died, probably of the Black Death, while a repeal of the ban against him was pending.

Ockham is a Schoolman of renown, but he marks the beginning of the end of scholasticism proper. By intent and avocation he was from first to last a theologian. By intellectual direction he was a sceptic, reminiscent of Hume. And by intellectual capacity he was a logician, and a most clear-thinking one too. His capacity for trimming away, as if with a razor's edge, logical redundancies and over-determinacies has become proverbial; and he would wield his razor in all contexts, logical, theological, and scientific. See, for instance, a selection of his philosophical writings by P. Boehner, 1957.

There are, on the face of it, traits of our modernity about

him. His logic, when formalized, is perhaps even more "modern" than that of Leibniz. In matters of sheer logical clarity and consistency he exceeded Galileo, and was not exceeded by Newton; and Kepler, alongside him, would appear to be outright irrational, most of the time at any rate. Also, Ockham is a leading person—always with a "modernity" of his own—in the physics of the peculiar 14th-century brand, which may be taken, or mistaken, for a pre-phase of the science of the 17th-century brand.

All this deepens the mystery of how it came about that there was such a hiatus in the advancement of physics in the 15th and 16th centuries. It is true that in the 16th century there were ample compensations, from astronomy, and also mathematics. But not so in the 15th century; it is wearying to have to fall back on the solitary name of Nicholas of Cusa, and to have to add something to his alleged greatness at each mention of him. And the rejoinder that in the arts, representational and literary, the 15th century was in no wise stagnant, makes the problem regarding physics the more acute.

OERSTED, CHRISTIAN (Rudkøbing 1777–Copenhagen 1851), Danish physicist and chemist; discovered, perhaps by chance, electromagnetism, that is, the effect of an electric current on a magnetic needle. As soon as the discovery was made public, there was many a physicist (Ampère, Arago, Poggendorff, Faraday, etc.) who felt that he should have really made it, and who could immediately enlarge upon and exploit the feat much better than Oersted was able to.

ORESME, NICOLE (Oresme, Normandie, 1325–Lisieux 1382), bishop of Lisieux; a great scholastic harbinger of the Renaissance; a leading scientist, and author of a tract on money; translated several books of Aristotle into French, especially his *Ethics,* and these may have been the first translations of Aristotle into any of the present-day "European" languages.

A full scale biography of Oresme is overdue, and badly needed.

ORPHEUS, a legendary Greek poet of Thracian origin. Plato, in a rash moment (*Cratylus* 402 B C), attributes to him the lines:

> Fair-flowing Ocean was the first to marry, and he
> wedded his sister Tethys, daughter of his mother.

This somehow immediately established Orpheus as an early cosmologist and philosopher; and only very courageous books on early Greek philosophy dare not to discourse on Orpheus profoundly and profusely.

PAPPUS OF ALEXANDRIA (c. 320 A.D.) would undoubtedly have become a first-rate research mathematician if he had lived at a time of mathematical spontaneity. But he lived at a time when mathematical creativity was at a low ebb, and all he could produce was a *Collection* of critical commentaries of, and of pseudocreative enlargements of, many achievements of others before him.

PARMENIDES OF ELEA (born c. 540 B.C.), great ontologist, creator of Being-as-Existence. Proper founder of Eleatic School, even if Plato suggests that Xenophanes before him ought to be so considered. Elea was in Southern Italy, not far from Paestum, of temple ruin fame.

Shakespeare's saying, through Hamlet,

> Thinking makes it so

pales when compared with the saying of Parmenides, 2,000 years before,

Thinking and Being refers to the same
to gar auto noein estin te kai einai

Parmenides posits a single Whole, which does not separate into physical, metaphysical, conceptual, or other Unities. This Whole, or One, is the same as the (Parmenidean) Being, and as the (Parmenidean) Existent; and it is not subject to growth and decay, but is immutable. Otherwise there would be Non-Being, but Non-Being is not thinkable.—Inexplicably, this One somehow has the spatial aspect of a (cosmic) sphere, and the sphere is homogeneous and bounded.

Toward the end of his essay Parmenides seems to soften his stand on Monism, and he apparently allows for some dualism of Light and Dark. Regrettably this end part of his doctrine is neither very intelligible nor very impressive.

In cursory reading, Parmenides, as also his near-contemporary Heraclitus, can be readily understood from knowing Homer, and "non-Homeric" words are even rarer in Parmenides than in Heraclitus. But, affectively, Heraclitus is wholly Homeric, even if much more somber in tone; whereas Parmenides, in difference from Homer, is very definitely "philosophical," even if outwardly the difference consists only in frequent occurrences of several forms of the Greek word for "to be."

Parmenides wrote in hexameter, not from mannerism but from convenience; Greek prose developed very slowly, and even the Attic prose of Thucydides is still quite uncouth. The lines in Parmenides are not composed of "wingèd words," but are harsh, jarring, and forbidding. But they are as taut, controlled, and compact, as the Parmenidean "Existent" which is enclosed in them; and they are as difficult to comprehend as Ontology itself always was and always will be.

PASCAL, BLAISE (Clermont-Ferrand 1623–Paris 1662), mathematician, physicist, philosopher. In science he achieved less that is specifically his, or is really forward-looking, than is sometimes assumed. For instance, his *probability* is still nothing more than *combinatorics,* without any anticipation at all of Bernoulli's law of large numbers, which was the actual debut of *probability* as a force in our civilization.

PERRAULT, CHARLES (Paris 1628–id. 1703, French writer. What made him famous was not so much the *Parallèles des anciens et des modernes* (1688–1698) as his *Contes de ma mère l'oye* (1697). The name was obviously derived from *Mother Goose's Melodies,* which appeared in London, England, 1760. But whereas the English book consisted of nursery rhymes, the book of Perrault consisted of stories of the type of *Little Red Riding-Hood.*

PESTALOZZI, JOHANN HEINRICH (Zurich 1746–Brugg 1827), Swiss pedagogue, "explicator" of Rousseau's *Émile,* founder of several "progressive" schools, author of "parent's guides" for "progressive education."

PETTY, SIR WILLIAM (Romsey, Hampshire, 1623–London 1687), English statistician and political economist; together with John Graunt, and with Jacob Bernoulli (through his *Ars Conjectandi*) he imparted to our present-day civilization the tendency to statistical enumeration of anything and everything.

Statistics *founded on probability* is perhaps the most exclusive characteristic of our civilization since 1600; and it would be difficult to find even a trace of it anywhere before.

PHILO OF ALEXANDRIA (13 B.C.–A.D. 54), tried to harmonize Faith and Reason, by showing, mainly through allegorical interpretations, that there are no incompatibilities between the (Jewish) Bible and (Platonic) Hellenism.

PLANCK, MAX KARL LUDWIG (Kiel 1858–Göttingen 1947), German physicist. He was a very expert thermodynamist, and the author of books that were read, but, for many years, not visibly more. Yet, academically he was advancing rapidly, and, as early as 1889, he was the successor to Kirchhoff at the University of Berlin. But suddenly, toward the end of 1900, at the age of 42, he introduced his quantum hypothesis for black body radiation; this was the true beginning of the 20th century, and the world has not been the same since.

Also, Planck may have been one of the first leading research physicists who are "theoretical" in the literal sense of not being themselves involved in the experimentation out of which their theoretization emerges.

The following is a mere chronology of certain events which culminated in the emergence of the quantum hypothesis. In 1859, G. Kirchhoff axiomatized the notion of a block body (or cavity), and he posited the existence of a universal radiation density function $F(\lambda, T)$. After that, for 20 long years nothing happened until, in 1879, J. Stefan suggested, as a guess from experimentation, that the all-wavelengths radiation density $u(T) = \int F(\lambda, T) d\lambda$ has the value $u = cT^4$, for some numerical constant c. After five years, in 1884, L. Boltzmann derived this value by daringly viewing radiation in a cavity as a thermodynamic gas à la Clausius, subject to the equation of state $p = \frac{1}{3} u$, which is Maxwell's equation for radiation pressure. In 1889, H. Hertz constructed his vibrator, and, in 1892, Willy Wien began to ruminate over its place in physics. But Wien made no use of it when deriving, in 1893, his radiation law that $T^{-5} F(\lambda, T)$ depends only on the product λT. In 1895, Planck began to interpret cavity radiation as energy of Hertzian origin, and five years later, in 1900, he thus produced his quantum hypothesis. And after another five years, in 1905, A. Einstein began to universalize this hypothesis by subjecting the photoelectric phenomenon to it.

BIOGRAPHICAL SKETCHES

PLATO OF ATHENS (427–347 B.C.) was the first, and most authentic, universalist of cognition and knowing. He was heir to Parmenides, Heraclitus, and Pythagoreans, one-time pupil of Cratylus, and full-time pupil of Socrates; and there would have been no Plato without them. But we, in turn, have inherited our intellectuality, directly or indirectly, from Plato; and without this we would not have the ability to discern what there was for Plato to inherit or learn from whomsoever he had done so.

Plato recognized, almost from the first, that there is an important structure of one-ness and sameness in the diverseness of our knowing, and that this one-ness unifies and links all our knowledge: the knowledge of the external world with the knowledge of our internal self; the knowledge of morals with the knowledge of mathematics; the knowledge of the individual with the knowledge of society; the knowledge about the schoolchild with the knowledge about the adult; the knowledge about a profession with the knowledge about poetry; the knowledge of what is right with the knowledge of what is holy; the knowledge about man with the knowledge about God.

Plato maintains, tirelessly, that the true texture of this one-ness is our rationality, whatever that be. There is only one way to knowing things, namely the rational way, and this way is neither selective nor restrictive; if it leads to knowing any one thing, it must lead to knowing every other thing. That is, there is only one single RATIO, take it, or leave it.

But Plato also saw the dilemma—and felt the anguish of the insight—that this very same rationality, however much it must be used to explain anything else, is entirely unable to explain its own self; rationality is not self-rationalizing, and the provenance of rationality is cloaked in an irrationality of its own. Thus, although even the knowing about God can only be arrived at rationally, yet, reversely, any rational knowing at all, about anything there is, must

[354]

somehow flow from a fountainhead of divinely inspired Supreme Intelligence; and Plato gave to this Intelligence an aspect of human-ness which he termed the Supreme Good.

Plato also knew—not very articulately, but sufficiently —that this dilemma also applies to, and is at its starkest in, mathematics. I think that by knowing this Plato had a certain superiority over mathematicians of his time; but there is no evidence for assuming that in matters of technical competence Plato was superior to, or even equal to, "professional" mathematicians in the Academy.

I think that philosophemes of Parmenides and Heraclitus made Plato also realize that it is difficult to separate the problem of Cognition from the problem of Re-cognition, and these two problems from the dilemma above. Plato somehow attempted to solve these problems and the dilemma, all in one, for better or worse, by the introduction of Ideas which are eternal, and of Souls which are immortal. The details of the solutions are inexhaustibly provocative, and the problems are as fresh, and as insoluble, today as on the day when Plato was first inspired to pose them.

Plato also had his pecularities.

From boyhood he dreamt of becoming not only a philosopher but also a "king," that is, a politician and statesman of consequence. From whatever reasons he never succeeded; probably because he simply did not have an aptitude for the game of politics. And in order to relieve his frustrations he wove into the *Republic* fantastic proposals which cannot but baffle the reader; and even in his old age, in the *Epistles,* he was still insisting that the world of politics had been in a conspiracy against him, and that he had been right and everybody else wrong.

Plato did not stop berating sophists. Whatever their faults, they did constitute the educational system of Greece, and Plato should have known that every educational systems always has shortcomings, and always needs reforming. And it is more effective to work hard at proposals for remedies than to harp on abuses.

Fortunately, Plato polished his written works more than any philosopher before or after him; and it is incongruous to have him reiterate that oral teaching is the primary need, and that putting things in writing is of secondary import only.

Finally, Plato's insistence on the theme of Eros and pederasty is misleading, and the presumed prevalence of the theme in Greek life and civilization is a near-invention of Plato's; there is almost nothing about the theme in Herodotus and Thucydides, in the extant plays of the tragedians, in Aristotle or other philosophers, or in the orators. Also, for instance in Homer, Achilles and Patroclus are comrades-in-arms and nothing else; and *Honi soit qui mal y pense.*

PLOTINUS (A.D. 205–270), THE neo-Platonist. His deity had a trinitary aspect, which had an influence on Church Fathers. It is difficult to be "neutral" about Plotinus. One either admires him or shrugs him off impatiently.

POINCARÉ, HENRI (Nancy 1854–Paris 1912), French mathematician on the level of Poisson and Cauchy. He founded combinatorial topology, and also the theory of automorphic functions, among other things.

He knew physics well; but his large textbooks in physics are not the measure of the man. He has quite a reputation in philosophy of mathematics and science, but in this area he is more articulate than profound.

POISSON, DENIS (Pithiviers 1781–Paris 1840), French mathematician, worked most successfully in virtually all parts of mathematics and mathematical physics. He was probably the greatest French mathematician in the 19th

century, but two circumstances conspire against this being
generally conceived of. First, he lacked a certain firmness
of assertion and conceptualization. Thus, while he incon-
testably was the creator of magneto- and electrostatics, yet
he did not set down the physico-mathematical conception
of a potential, but he speaks colorlessly of "l'expression."
And secondly, there is no edition of his "Oeuvres
complètes" to impress us with the volume and magnitude
of his achievements.

Polybius (Megalopolis 200–id. 125 B.C.), a very important historian. His work is written with the artlessness,
and even dullness of form, which is found in the works of
"professional" historians, who are afraid of being attentive
to style lest they be distracted by this from concentrating
exclusively on "objectivity" of content.

Porphyry (Tyr 234–Rome 305); "typical" neo-Platonist;
also, by chance, violently anti-Christian. Among his "notorious" works are a *Life of Plotinus* (which begins with
the statement that Plotinus was ashamed of having to have
a body) and a *Life of Pythagoras*. Of historical consequence was a work of his on Aristotelian categories; a
Latin translation by Boëtius was, in the Middle Ages, a
much used compendium of Aristotelian Logic.

Posidonius (Apamea, Syria 135–Rome 50 B.C.), an
articulate representative of the Middle Stoa; very popular
with highly placed Romans like Cicero and Pompey. He
was versatile and knowledgeable; and his main work,
apparently, was a historical work, now lost, which was
much used by others.
    Since the turn of the century, scholars have been
drawing an ever larger portrait of him, so much so, that to
do no more than compare him to Aristotle, say, is
considered to be an understatement by some. The Chief

Exaggerator is Karl Reinhardt. The latter was originally a member of the circle around the poet Stefan George, and the historians of that circle all liked the "heroes" of their manufacturing to be "colossal."

For a confirmation of the assertion that the historian-members of the circle around Stefan George were wont to draw their biographees in "heroic" dimensions, see the entry on Ernst Kantorowicz and his *Kaiser Friedrich der Zweite* in the encyclopedia *Der grosse Brockhaus,* 16th edition (1955), VI, 272.

POYNTING, JOHN HENRY (Monton 1852–Birmingham 1914), English physicist. His main achievement was not the introduction of the "Poynting vector," but the measurement of the average density of the Earth (1890).

PRIESTLEY, JOSEPH (Fieldhead near Leeds, 1733–Northumberland, Pennsylvania, 1804), a leading Anglo-American chemist; discoverer of oxygen.

He was a known personage; preacher, philosopher, liberal.

PROCLUS (Constantinople 412–Athens A.D. 485), the "Divine," was a "high-ranking" neo-Platonist; composed commentaries on Platonic Dialogues, of which several are extant, a book on theology, and, very significantly, a commentary on Book 1 of Euclid, with two introductions.

The commentary is not at all a "great" book; but it is the only one of its kind, and the information which it transmits is very important to the history of mathematics.

PTOLEMY (c. 90–A.D. 168), or Claudius Ptolemaeus, composed, in addition to his majestic treatise on astronomy, also an equally celebrated *Guide to Geography,* which, together with its maps (of which some may not have been his own), played a great role in the history of discoveries in the 16th century.

An *Optics* of his, which only exists in a Latin translation from the Arabic version, has lately been receiving attention in very detailed studies.

PYTHAGORAS AND PYTHAGOREANS. Pythagoras of Samos lived in the second half of the 6th century B.C.; he founded an important community in Southern Italy, and, especially in Aristotle, "the Italians" is virtually a synonym for "the Pythagoreans." To judge by late reports about him, most of which are adulatory, he must have achieved the nearly impossible feat of being, in one and the same person, a searching scientist-philosopher; a convincingly holy man; a consummate politician and statesman; and his own best publicity man, even if crowds of followers, mostly naïve ones, were tirelessly spreading their belief in his sainthood and superiority.

There must have been a Pythagoras, and one of some stature. In the 5th century B.C., Herodotus (iv, 95) makes mention of one, and in a rather respectful manner. And even before that Heraclitus gives Pythagoras the recognition of bracketing him with Hesiod, Xenophanes, and Hecataeus in an ill-tempered outburst (Heraclitus, fragment 16) that each of them is not, as apparently reputed, a polymath but is, in truth, an ignoramus.

However, Plato and Aristotle are most reluctant to speak of Pythagoras, as if, amid the welter of facts and fancy about him, they felt uncertain about his true identity and achievements. Plato says hardly anything even about Pythagoreans in general. Aristotle however, most fortunately, does speak about them; and we will quote part of a famous passage about them from the *Metaphysica,* Book I, Chapter 5, in which Aristotle describes, and interprets, how experience in musical acoustics engendered in the minds of Pythagoreans the expectation that there might be a comprehensive system of physical knowledge which is constructed entirely out of mathematics.

Contemporaneously with these developments [in non-mathematical physics] and even earlier, the so-called Pythagoreans were attaching themselves to mathematics; they were the first to promote it and they engrossed themselves in it; and they came to believe that its principles are the principles of all things. Since among the principles of mathematics numbers are the leading ones, the Pythagoreans seemed to discern in them many likenesses to things as they are, or come to be; so that, such and such a modification of number presents *justice,* another one presents *soul* or *mind,* still another one *opportunity,* and similarly, more or less, with all the rest. They furthermore recognized that the attributes and ratios of musical scales ("harmonies") reside in numbers; and it revealed itself to them that all other things are modeled after numbers, and that numbers are the primary objects in the whole of nature; they therefore made the assumption that the elements of numbers are the elements of all things and that the entire universe is a 'harmony' and a 'number.' And whatever similitudes there appeared to be between numbers and scales on the one hand, and the attributes, parts, and arrangement of the universe on the other hand, these they collected and correlated. And if there was a gap anywhere, they readily devised an innovation by which to make the system a completed whole. For instance, on the presumption that the number 10 is something perfect which comprises the whole essential nature of the numerical system, they assert that the bodies which move in the heavens are ten; but, since there are only nine that are visible, they posit an anti-earth as the tenth."

May I add on my own that every part of this exposition is a prophecy. Theoretical physics constantly transfers mathematical paradigms from one area of physics to another, and generalizes them from smaller areas into larger ones. And if one is willing to concede that the so-called linear oscillator of present-day physics is of acoustical origin, then even the Pythagorean presumption that physics proper is an unfolding of acoustics retains a measure of validity.

The Pythagorean assigning of integer numbers to basic notions like *justice, soul, opportunity,* etc., has blossomed out, in the "sophisticated" mathematical logic of today, into assigning integer numbers to formulas in a formal system (Gödel numbers).

And the Pythagorean invention of an anti-earth because of some preconception of theirs about the mathematical nature of the number 10 has matured into the introduction of anti-particles for the sake of filling out some gap or gaps in the mathematical theory of the particles. The theoretical rationale for the introduction of the anti-electron (that is, positron)—which was the first of its kind—was the purely mathematical fact that the relativistic equations of the electron remain unchanged if one replaces the value $+e$ of the electron by its negative value $-e$.

Of course, Aristotle did not at all know our present-day physics, not even by the faintest or remotest kind of divinatory anticipation. But this does in no wise invalidate our privilege to rationalize, against the background of our own experience, the explanations that were given by Aristotle against the background of the totally different experience of his own age and person. For it is the mark of the greatness of a historian that he pronounces truths that will endure, and ring out, long after his own background, against which he uttered them, has crumbled away and turned into the dust out of which later backgrounds will continue to be built.

RANKE, LEOPOLD VON (Wiehe, Thuringia, 1795–Berlin 1886) was a historian's historian.

He drew up Rules of Historiographic Conduct, evident ones, which are predicated on the presumption that there is an absolute Truth, and that it can be arrived at if properly searched for. This presumption is almost inapplicable to history of "ideas"; even if it be applicable to history of "facts," if there are any "facts" which are divorced from "ideas."

RICCI, CURBASTRO GREGORIO (Lugo, Ravenna, 1853–Bologna 1925), Italian mathematician, created, knowingly, for the future, a basic technique, which is named after him.

RIEMANN, BERNHARD (Breselenz, Hannover, 1826–Selasca sur le lac Majeur, 1866), an extraordinary mathematician. His main strength lay in his capacity to erect geometric structures for the purposes of analysis, but he could also master topics which were at a distance from this approach.

On an occasion, he boldly envisaged, in print, the possibility of a theory of physics in which all data, including the underlying space, would be numerically discrete (that is, "quantized").

ROMER, OLE (Aarhus 1644–Copenhagen 1710), Danish astronomer, determined in 1676, the velocity of light by observing the eclipses of the satellites of Jupiter. He was also a public figure.

RUIZ, JUAN (died apparently, in 1350), known as the "archpriest of Hija," was the Chaucer of Spain, only, perhaps, even "naughtier."

SCHIAPARELLI, GIOVANNI VIRGINIO (Savigliano 1835–Milan 1910), enterprising astronomer. In 1861 he discovered the asteroid Hesperia, in 1877 he proposed a reconstruction of the planetary system of Eudoxus, and, alas, in 1887 he threw out the suggestion that there are *canali* on the planet Mars.

SEXTUS EMPIRICUS (c. A.D. 190), physician and philosopher, lived in Alexandria and Athens. In philosophy he was a leading "Sceptic," and two lengthy works of his survive. They are boring, but important. For instance, the prooemium in the poem of Parmenides comes from Sextus.

SOMMERFELD, ARNOLD (Königsberg 1868–Munich 1951), German physicist. He contributed greatly to the advancement of physics by writing very lucid textbooks.

SOPHOCLES (495–406 B.C.), perhaps the "stagiest" of the three Greek Tragedians, master of the "tragic irony" in which events pervert the intent of a verbal statement. As a person, Sophocles was reputed to be most gentle, but at his "writing desk" he dealt in cruel fate. His *Oedipus King* has become a household word since Sigmund Freud named his "Oedipus complex" after him; but the parallel "Electra complex" as a name of sister-for-brother love has not taken root. Antigone, heroine of *Antigone,* is the lovably intractable "sweet maiden"; and the contrast between her and her "conformist" sister Ismene is timeless.

SPENGLER, OSWALD (Blankenbarg, Harr, 1880–Munich 1936), author of the *Decline of the West*. His views on the role of mathematics as a characteristic of (the Spenglerian) "culture" are always interesting and frequently correct, provided one makes allowances for the fact that his book is the work of a "dilettante" and not of an academic "professional."

[363]

STEVIN, SIMON (Bruges 1548–The Hague 1620), the first recognizably "modern" physicist, who somehow sounds very different from the physicists of the Middle Ages, although in the main he is working in statics only. He resumed hydrostatics where Archimedes had left off, and he made the first attempts to advance beyond him.

STONEY, GEORGE JOHNSTON (Oakley Park, Kings' County, 1826–London 1911), Irish astronomer and physicist, was one of the first creators of the modern conception of an electron.

TANNERY, PAUL (Nantes 1843–Pautin, Seine, 1904), notable Greek scholar, with the mathematical training of a "polytechnicien"; specialized in history of Greek science; he sometimes over-speculates when documentation is sparse.

THALES OF MILETUS (c. 624–547 B.C.), a dominant figure in textbooks on the history of Greek philosophy. In the main, this dominance was created by Aristotle, who lived 250 years after Thales. References to Thales in Herodotus, Aristophanes, and Plato, all before Aristotle, are not incompatible with the Thales of Aristotle; but on their own they do not suggest a Thales like his. On the other hand, after Aristotle, there are in Proclus, with hints in Diogenes Laertius, late but specific reports which make Thales into a mathematician of consequence; and Proclus asserts that these reports are derived from Eudemus, the pupil of Aristotle. However, there is nothing about a Thales of this kind in Aristotle himself, or in any of the numerous passages about Thales in the doxographic collection of H. Diels. Undoubtedly there was a Thales. But there is also a problem of who and what he was; and, due to the historiographic interest of the problem, we will enlarge upon it.

In a lengthy passage in the *Metaphysica* (Book I, 983 b 7–984 a 5), Aristotle lays down the thesis that Thales was more or less the founder of natural philosophy in general, and the founder of hylozoist monism in particular; and he asserts that the generative substance of Thales was Water. Next, this last assertion is repeated in *De Caelo* (294 a 28–33). Next (*Ethics,* 1141 b 2), Thales has a typical philosopher's impracticality in everyday matters. However (*Politics,* 1259 a 9), when businessmen taunt him with his poverty, which is due to the uselessness of philosophy, Thales teaches them an object lesson by contriving a business operation which is very costly to them. Finally (*De Anima,* 405 a 9–22, 411 a 7–9), Aristotle speculatively attributes to Thales two utterances, of secondary import, from the area of "animation of matter" which go well with his brand of Monism.

About 100 years before Aristotle, in Herodotus (*Histories,* Book I, Chapters 74, 75, 170), Thales is an "activist" and not a "philosopher," and in no reported detail does he overlap with the Thales of Aristotle. The Thales of Herodotus has the astronomical knowledge for predicting an eclipse to within a year; he has the engineering knowledge for proposing a method by which an army might ford a river over which there are no bridges; and he has the statesman's knowledge for making constructive suggestions in matters of foreign policy. Also, the Thales of Herodotus is not a figure of the heroic past, or a "sage" like Solon, or an introverted brooder like some of the priests whom Herodotus must have met at Egyptian shrines, but a very normal person with exceedingly valuable qualifications.

After Herodotus, in Aristophanes, about 60 years before Aristotle, Thales is a "popular" name of somebody in the past, who is reputed to have possessed technical competences on a prodigious scale; twice (line 170 of the *Clouds,* and line 1005 of the *Birds*) there are, for comic effect, exclamations of the kind: "This fellow is a veritable Thales!" However, from the extant allusions in Aristophanes it would not be easy to decide whether, as suggested by George Sarton (*A History of Science. Ancient Science through the Golden Age of Greece,* 1952, p. 169), the Thales of popular fancy was a Benjamin Franklin, who, although a public figure of political status, could also contrive stoves and lightning rods; or whether this Thales was a lifelong producer of gadgets, like Thomas Edison; or whether, like the Albert Einstein of popular fancy, he was endowed with some occult mathematical powers with which, by a sleight-of-hand or, rather, sleight-of-mind, to produce intricate levers, or pulleys (or bombs), and suchlike objects.

Plato corroborates this side of Thales in the *Republic,* Book X, 600 A; but Aristotle again not. Next, in Plato (*Protagoras,* 343 A-B) Thales is—what he was not in Herodotus—a sage, in company with other sages, and he is co-author of pithy sayings on the level of: "Know thyself," and "Nothing overmuch." Finally, in *Theaetetus* 174 A-B, Thales is an absentminded astronomer who, when "star-gazing" while walking, falls into a pit, to the amusement of a saucy wench who passes by. The Thales of this anecdote has a "family resemblance" to the Thales of Aristotle. It must be stated, however, that nowhere in Plato is there a trace of an awareness of hylozoist monism, or of an Ionic school, and Anaximander and Anaximenes are never mentioned, or alluded to. The only Monism of which Plato is aware is the "noetic" monism of the "Eleatic sect" (Ἐλεατικὸν ἔθνος), and the "totemistic" ancestor of it is Xenophanes (Sophist, 242 C-E).

Several centuries after Aristotle, in Diogenes Laertius, a note of uncertainty about the role of Thales crept into his introductory report about the origins of philosophy. In I, 14, Diogenes reports (all quotations are from the Loeb translation) that "the succession passes from Thales through Anaximander, Anaximenes, Anaxagoras, Archelaus, to Socrates," which agrees with the thesis of Aristotle; but in the preceding sentence I, 13, Diogenes somehow seems to weaken the role of Thales by putting it thus: ". . . philosophy, the pursuit of wisdom, has had a two-fold origin; it started with Anaximander on the one hand, with Pythagoras on the other. The former was a pupil of Thales, Pythagoras was taught by Pherecydes."

Another uncertainty, a harmless one, is the doubt in Diogenes (I, 35) whether it was Thales or Pythagoras, who "was the first to inscribe a right-angled triangle in a circle."

Finally, Diogenes (I, 28) has no lingering doubts when reporting that "Hieronymus informs us that he [Thales] measured the height of the pyramids by the shadow they cast, taking the observation at the hour when our shadow is of the same length as ourselves."

This last information is of the same texture as later reports in Proclus, who, in his Euclid commentary, credits Thales with the following discoveries: that a diameter halves a circle; that in an isosceles triangle the angles at the base are equal (except that Thales, archaically, called the angles "similar"); that opposite angles are equal; and that triangles are congruent if a side and two adjacent angles are equal.

These achievements, however "elementary," if indeed accomplished in the 6th century B.C., would indicate that Greek mathematics from its very beginnings had the tendency to be self-reflective, and to be "theoretical" in a manner which is different from the manner in which Egyptian and Babylonian mathematics had been shaping themselves. But the absence of corroboration in Plato and Aristotle that Thales had been involved in suchlike achievements casts a shadow of uncertainty over the late reports that he had so been.

All told I would say that there are as many Thaleses as

there are separate sources of information about him, and that the Thales of textbooks is the Thales of Aristotle, with the other Thaleses somehow superimposed.

THEOPHRASTUS OF ERESUS (372–287 B.C.)´, disciple first of Plato, then of Aristotle; successor and "heir" to Aristotle in the *Peripatetic School*. He was an arch systematizer and classifier of all knowledge according to its history and structure; a list of publications of his was 400 titles strong. He was a most knowledgeable naturalist; but he could discourse and write on literally anything on earth and under the sun. He was also a man of the world, very witty and very popular, even with the shade of an "intellectual" matinee idol about him.

But there was not a trace of a hint of the genius of Aristotle in him, and nothing about him is irritating because all about him is soporific.

THOMPSON, D'ARCY WENTWORTH (Edinburgh 1860–St. Andrews 1948), was a zoologically oriented biologist, and a classical scholar. His remarkable book *On Growth and Form* is a nostalgic survey of 19th-century biology as it still then was; and he also made, among other works of scholarship, an annotated translation of Aristotle's *Historia Animalium*.

On Growth and Form is not for a reader wanting to know the latest about detection of physiological rhythms, composition of cellular acids with tongue-tripping names, stringing of ladders of molecules, deciphering of combinatorial codes, etc. But this post-Edwardian opus is a treasure house of biological variations on themes of design, pattern, structure, and symmetry, and the variations are as enchanting as they are meaningful.

VOLTAIRE (Paris 1694–id. 1778). There is probably no "writer" nowadays who could interpret a theory of physics of the last 50 years as competently as Voltaire did Newton's gravitation.

WEIERSTRASS, KARL (Ostenfelde, Westphalie, 1815–Berlin 1897), German mathematician.

He liked his Rhine wine rich; but this somehow did not prevent him from becoming a leading enemy of fuzziness in mathematical thinking in the 19th century.

WEYL, HERMANN (Elmshorn 1885–Zurich 1955), prominent mathematician and connoisseur of physics. His distinctiveness came through better in his books, even synthesizing ones, than in his memoirs in periodicals; which is rare in mathematics.

WHEWELL, WILLIAM (Lancaster 1794–Cambridge 1866), was very prolific. In addition to his large works on history and philosophy of inductive sciences, he also had many works on mineralogy, mechanics, and mathematics. And he even had a book on the architecture of German churches, in addition to articles and dissertations on poetry, say.

WHITEHEAD, ALFRED NORTH, the best philosopher of mathematics in the first half of the 20th century; except that for him mathematics was some variety of logic, which it is not. Nobody will deny, from whatever approach, that Aristotle was a great logician; and yet, his sense for the mystique of mathematics was very limited, and certainly much less authentic than in the case of logic.

WITELO (c. 1230–c. 1275), composed, among other works, an optical treatise *Perspectiva,* which enlarged upon the work of Alhazen. A book of Kepler's in 1604 is entitled *Ad Vitellionem Paralipomena,* that is, *"Leftover Additions" to Witelo.*

The biographical data of Witelo are somewhat uncertain. He was born near Lagnica (= Liegnitz) in Silesia, of a Thuringian father and a Polish mother, and died in a Premonstratensian House at Witów (Vitonia) near the historical city of Piotrków Trybunalski in Poland. He studied in Padua; completed, in 1270, his *Perspectiva* in Viterbo, then the seat of the Roman Curia; and dedicated the work to a fellow resident, William of Moerbecke, then penitencer to the Church. The work was printed in Nuremberg in 1535 and, by Frederic Risner, in Basel in 1572.

XENOPHANES OF COLOPHON (570–475 B.C.), famed poet, longlived, itinerant, self-pitying; with a strong bend towards theology and philosophy.

His general wisdom is exemplified by his saying that if cattle or lions had hands, with which to paint and produce works of art, the horses would paint their gods like horses, and the cattle like cattle. Xenophanes attacked the religion of Homer and Hesiod for the anthropomorphisms and immoralities of its gods. He inclined towards a monotheism; but when he extended his monism to physics, the result was simplistic. Such was, at any rate, the view of Aristotle, as can be seen from the following summary judgment in *Metaphysica* I, 5, 986 b 23. "Xenophanes first taught the unity of things, Parmenides is said to have been his pupil, but he did not make anything clear, nor did he seem to get at the nature of things, but looking up into the broad heavens he said: The unity is god." (Translation from Arthur Fairbanks, *The First Philosophers of Greece*, 1898, p. 79.)

Xenophanes has been over-praised for the simple observation that when shells or "imprints" of sea animals are found inland, then that part of the land must have once been covered by the sea.

[370]

ZELLER, EDUARD (Kleinbottwarr, Wurtemberg, 1814– Stuttgart 1908), author of a monumental history of Greek philosophy, all parts of which continue to be indispensable and unsupersedable.

ZENO OF ELEA (flourished c. 460 B.C.), fierce defender of his teacher Parmenides against pluralists; fierce defender of political liberty against a tyrant, who then tortured him to death.

Generations of freshmen in philosophy have thrilled to the ever-fresh tales of the ever-intractable puzzles of Zeno. Simplicius transmitted several important puzzles, about One And Many, even in Zeno's own words, but they have not become popular. The memorable puzzles are those preserved by Aristotle in his *Physica*. They are about motion; and "Achilles" and "the flying arrow" are the leading ones: (i) Achilles cannot overtake a turtle, since whenever he reaches the point from which the turtle started the turtle has by then a new lead. (ii) If time is composed of moments, a would-be flying arrow must at each moment be at rest, and thus can never fly.

The puzzles against motion cannot be refuted on Zeno's home ground, that is, in Zeno's own mode of verbalization. The concept of motion must first be "formalized," and it suffices that it be put into the setting of the Euler–Lagrange mechanics. In this setting the puzzles cease to be formulable.

INDEX